Essays in Biochemistry

Other recent titles in the Essays in Biochemistry series:

Essays in Biochemistry volume 38: Proteases in Biology and Medicine
edited by N.M. Hooper
2002 ISBN 1 85578 147 6

Essays in Biochemistry volume 37: Regulation of Gene Expression
edited by K.E. Chapman and S. J. Higgins
2001 ISBN 1 85578 138 7

Essays in Biochemistry volume 36: Molecular Trafficking
edited by P. Bernstein
2000 ISBN 1 85578 131 X

Essays in Biochemistry volume 35: Molecular Motors
edited by G. Banting and S.J. Higgins
2000 ISBN 1 85578 103 4

Essays in Biochemistry volume 34: Metalloproteins
edited by D.P. Ballou
1999 ISBN 1 85578 106 9

Essays in Biochemistry volume 33: Molecular Biology of the Brain
edited by S.J. Higgins
1998 ISBN 1 85578 086 0

Essays in Biochemistry volume 32: Cell Signalling
edited by D. Bowles
1997 ISBN 1 85578 071 2

volume 39 2003

Essays in Biochemistry

Programmed Cell Death

Edited by T.G. Cotter

Series Editor
T.G. Cotter (Ireland)

Advisory board
G. Banting (U.K.)
E. Blair (U.K.)
C. Cooper (U.K.)
P. Ferré (France)
J. Gallagher (U.K.)
N. Hooper (U.K.)
J. Pearson (U. K.)
S. Shears (U.S.A.)
D. Sheehan (Ireland)
J. Tavare (U.K.)

Portland Press

Essays in Biochemistry is published by Portland Press Ltd
on behalf of the Biochemical Society.

Portland Press
59 Portland Place
London W1B 1QW, U.K.
Fax: 020 7323 1136;
e-mail: editorial@portlandpress.com
www.portlandpress.com

British Library Cataloguing-in-Publication Data
A catalogue record for this book is available from the British Library

ISBN   1 85578 148 4
ISSN   0071 1365

Typeset by Portland Press Ltd
Printed in Great Britain by Black Bear Press Ltd, Cambridge

Contents

3   Caspases: the enzymes of death
Boris Zhivotovsky

4   Apoptosis: bombarding the mitochondria
Philippe Parone, Muriel Priault, Dominic James, Steven F. Nothwehr and Jean-Claude Martinou

5   Death receptors
Harald Wajant

6   Guardians of cell death: the Bcl-2 family proteins
Peter T. Daniel, Klaus Schulze-Osthoff, Claus Belka and Dilek Güner

7 Oncogenes as regulators of apoptosis

Mohamed Labazi and Andrew C. Phillips

8 The final step in programmed cell death: phagocytes carry apoptotic cells to the grave

Aimee M. deCathelineau and Peter M. Henson

9 Apoptosis in disease: about shortage and excess
Thomas Brunner and Christoph Mueller

10 Therapeutic approaches to the modulation of apoptosis
Finbarr J. Murphy, Liam T. Seery and Ian Hayes

11 Apoptosis: future perspectives
*Carol Ward, Adriano G. Rossi, Christopher Haslett
and Ian Dransfield*

Preface

Life and death, even at the level of the cell, are but opposite sides of the same coin. Just as the cell-generating process of mitosis is tightly regulated at the genetic and molecular level, so too is the cell-death process of apoptosis. Much of our understanding of programmed cell death, or apoptosis, of cells has come from a deluge of research output during the last decade of the 20th century. However, there are still considerable gaps in our knowledge and I expect that these will be filled slowly over the coming years. The excitement generated by research in the field stems mainly from the broad implications of our understanding of how the process is regulated, and how it may be manipulated in diseases and in developmental and commercial processes. The timely award of the Nobel Prize to Robert Horvitz from his seminal work on the regulation of programmed cell death in *Caenorhabditis elegans* indicates a certain maturity in the field.

I hope that this volume of essays, which covers most of the exciting areas of research in the field of apoptosis, will provide a snapshot of our knowledge at the beginning of the 21st Century. The volume begins with a historical view of the field from my own laboratory indicating when key discoveries were made and how they subsequently influenced the development of the field. This is followed by chapters from Justin McCarthy and Boris Zhivotovsky who explore the role of apoptosis in development and the function of the caspase enzyme cascade, both areas of sustained interest at present. Philippe Parone and colleagues document the role of the mitochondrion as a pivotal point in the execution phase of apoptosis and discuss the Jekyll-and-Hyde-type role played by cytochrome *c*. The chapter by Harald Wajant takes a look at the initiation phase of the whole process and the roles played by death receptors, in particular members of the tumour necrosis factor family.

Peter Daniel and colleagues look at the enigmatic Bcl-2 family of proteins and the controlling role they play in apoptosis. This is followed by three chapters from the laboratories of Andrew Phillips, Peter Henson and Thomas Brunner on the central role played by apoptosis in a frightening array of disease processes. In the Henson chapter, the focus is on the how the dead cells are removed, which in the scheme of things, is the last piece of the apoptosis jigsaw. The final two chapters from the Ian Hayes and Ian Dransfield laboratories attempt quite elegantly to peer into the future and see what it holds for this exciting field of research.

With a short volume of this sort on a subject as broad as apoptosis, there will invariably be gaps, but if the interest of the reader is sufficiently stimulated by what has been written, then perhaps these gaps can be filled by further reading in this exciting area of science. Material will not be a limitation!

Finally, my thanks go not only to the authors of each essay, but also to the staff of Portland Press, and in particular, Rhonda Oliver and Mike Cunningham who have steered me through the whole process. It has been an education!

Tom Cotter
Cork, Ireland,
July 2003

Authors

Tom Cotter is Professor and Chair of Biochemistry in University College Cork (Ireland). He graduated with a D.Phil. from the University of Oxford and studied as a postdoctoral fellow in Colorado, California and Germany before moving back to Ireland. He became interested in the field of apoptosis in 1990 and has worked to understand the role played by apoptosis in disease ever since. At present, he is the principle investigator of a large research group that studies apoptosis-inducing signalling in cancer and neurodegenerative disorders. **James Curtin** obtained a B.Sc. in Biochemistry at University College Cork in 1999. He stayed in Cork to study Fas-receptor-mediated apoptosis in prostate cancer under the supervision of Tom Cotter and was awarded his Ph.D. in Biochemistry recently. James has accepted a postdoctoral fellowship in the Gene Therapeutics Research Institute, Los Angeles, CA. Here he will study the potential of gene therapy directed against brain tumours and neurodegenerative disorders.

Justin McCarthy obtained a Ph.D. in Biochemistry from University College Cork (Ireland) in 1996. He then undertook postdoctoral research on the cloning and characterization of genes and proteins involved in apoptosis in the laboratory of Vishva M. Dixit at the University of Michigan and Genentech Inc., South San Francisco, CA. Subsequently, he held a position as Senior Scientist at a biotechnology company, Scios Inc., Sunnyvale, CA. He was appointed as a Lecturer in Biochemistry at University College Cork in 2002. His present research interests involve the Presenilin genes and their involvement in neurodegeneration and the pathogenesis and progression of Alzheimer's disease.

Boris Zhivotovsky received his Ph.D. in Biochemistry and Radiobiology in 1975 and his Dr. Sci. in 1989 in St. Petersburg (Russia). In 1991, he joined the group of Sten Orrenius at the Karolinska Institutet, Stockholm, Sweden, where he was later appointed Professor of Toxicology. His initial work on mechanisms of radiation-induced lymphoid cell death led to continued interest in understanding how radiation kills cells. At present, his general research interests are focused on cell death mechanisms of importance for the elimination of cancer cells; in particular, on deficiencies in apoptotic machinery of tumour cells resistant to treatment. His group made a notable contribution to characterization of caspases and their localization and translocation during apoptosis.

Philippe Parone obtained his doctoral thesis from the University of Cambridge (U.K.), following work on the role of the interaction between Bcl-2 family members in the regulation of neuronal apoptosis. He is currently investigating the importance of mitochondrial fission in apoptosis and cell-cycle regulation. **Muriel Priault** obtained her doctoral thesis from the University of Bordeaux (France) following work on mitochondria and apoptosis, and, in the meantime, benefitted from a training in electrophysiology at the New York University. She then joined Jean-Claude Martinou's laboratory where she works on the role of mitochondria in autophagy. **Dominic James** initially worked at EMBL exploring limb-pattern formation and the role of fibroblast growth factor-2 in embryonic development. His research focus shifted towards apoptosis and he undertook a Ph.D. at Manchester University (U.K.) studying the molecular mechanisms of apoptosis following growth factor withdrawal in blood cells. His work currently involves investigating the impact of mitochondrial fission on apoptotic pathways. **Steve Nothwehr** performed graduate work at Washington University in the laboratory of Jeffrey I. Gordon, where he studied signal peptidase processing of secretory proteins. After postdoctoral work with Tom H. Stevens at the University of Oregon he started his own laboratory at the University of Missouri where he has studied membrane trafficking in yeast. He is currently on research leave in the laboratory of Jean-Claude Martinou at the University of Geneva. **Jean-Claude Martinou** is Professor in the Department of Cell Biology at the University of Geneva. He is interested in the mechanisms of action of Bcl-2 family members and the role of mitochondria in cell death and in the processes that underlie mitochondrial fission and fusion.

Harald Wajant is Full Professor and Head of the Department of Internal Molecular Medicine at the Medical Polyclinic of the University of Wuerzburg (Germany). He received his Ph.D. degree in Biology from the University of Stuttgart in 1993. His current research interests are in the areas of apoptotic and non-apoptotic signalling by death receptors, and the development of death-ligand derivatives with cell-surface target-restricted activity for cancer treatment.

Peter Daniel was born on 16 December 1960 in Reutlingen, Germany. He undertook medical studies at the Eberhard Karls University in Tübingen, Germany, after which he became a post-doc at the Institute for Tumor Immunology, German Cancer Research Center, Heidelberg. He is a consultant in haematology and oncology. At present, he is group leader of the Molecular Hematology and Oncology Section at the University Medical Center Charité, Humboldt University, Berlin, Germany. His research fosuses on the role of the CD95 death receptor in immune homoeostasis; the identification of defects in caspases and pro-apoptotic Bcl-2 family members in cancer; functional genomic analysis of apoptosis defects in human malignant tumours and their relevance to disease prognosis and the predictive value for responses to anti-cancer therapy. **Klaus Schulze-Osthoff** was born on 26 August 1960 in Münster. Germany. He studied biochemistry at the Westphalian Wilhelms University in Münster and became a post-doc at the Vlaams Institute of Biotechnology, Gent, Belgium. He was a group leader at the Institute for Tumor Immunology, German Cancer Research Center in Heidelberg and at the Institute for Biochemistry, University of Freiburg, Germany. He also held the post of Associate Professor at the Universities of Tübingen and Münster. At presently, he is Director of the Institute for Molecular Medicine at the Heinrich Heine University, Düsseldorf, Germany. His research focuses on death-receptor signalling and the roles of caspases in apoptosis. **Claus Belka** was born on 15 May 1967 in Gelsenkirchen, Germany. He studied medicine at the University of Essen. He is a consultant in radiation oncology and group leader of Experimental Radiation Oncology Section at the Radiation Oncology Department of the Eberhard Karls University of Tübingen, Germany. Experimental work focuses on the role of Bcl-2 family members in radiation-induced apoptosis and the role of the endoplasmic reticulum. **Dilek Güner** was born on 2 November 1969 in Eskeşehir, Turkey. She studied medicine at Hannover University and is now a specialist in radiation oncology at the Department of Radiation Oncology, University Medical Center Charité, Humboldt University, Berlin, Germany. Her research deals with apoptosis defects and Rb pathway cell-cycle deregulation in human malignant tumors and their relevance to disease prognosis and the predictive value for responses to radiation therapy.

Mohamed Labazi obtained a B.Sc. from the My Ismail University (Morocco), a Masters Degree from the St. Petersberg State Technical University, Russia and a Ph.D. from the University of Barcelona, Spain. After post-doctoral training with William S. Dynan, he took up a position as a research fellow in laboratory the laboratory of Andrew Phillips at the Institute of Molecular Medicine and Genetics, Augusta, GA. **Andrew C. Phillips** studied for a B.Sc. in genetics at the University of Nottingham (U.K.) and completed his Ph.D. at the Beatson Institute for Cancer Research in Glasgow (U.K.) After post-doctoral training with Karen H. Vousden at the Ludwig Institute, London and at the National Cancer Institute, Frederick, U.S.A., he took up a position as an Assistant Professor at the Institute of Molecular Medicine and Genetics, Augusta, GA. He is a Georgia Cancer Coalition Distinguished Scholar.

Aimee M. deCathelineau received her B.Sc. in Genetics and Cell Biology from the University of Minnesota, Twin Cities, MN, and her Ph.D. in Experimental Pathology from the University of Colorado Health Sciences in Denver, CO. She is currently a postdoctoral fellow with Dr Gary Bokoch at the Scripps Research Institute in La Jolla, CA. **Peter M. Henson** received his Veterinary degree from the University of Edinburgh and a Ph.D. in Immunology from the University of Cambridge. He spent 10 years as first a postdoctoral fellow and then faculty member at the Scripps Clinic and Research Foundation in La Jolla, CA, and since 1977 has been on the Faculty of the National Jewish Medical and Research Center and the University of Colorado Health Sciences Center in Denver, CO.

Thomas Brunner is Assistant Professor in Experimental Pathology at the Institute of Pathology, University of Bern, Switzerland. He has a long-standing research interest in the regulation of T-cell apoptosis and cytotoxicity in the pathogenesis of inflammatory diseases. **Christoph Mueller** is Associate Professor at the Institute of Pathology, University of Bern, Switzerland. His research focus has been in various fields of immunopathology, including the role of cell-mediated cytotoxicity in transplant rejection, auto-immune diabetes and inflammatory bowel disease.

Finbarr J. Murphy is Head of Apoptosis Biology at EiRx Therapeutics, an apoptosis drug discovery company based in Cork (Ireland). Before taking up a position at EiRx, he held a number of postdoctoral positions both in Ireland and the U.S.A., including an ORISE fellowship at the Food and Drug Administration, Bethesda, MD, where he examined the regulation of interleukin-12 secretion. He holds a Ph.D. from University College Dublin in the area of inflammation biology. His current research interests are survival pathways in inflammatory and transformed cells. **Liam T. Seery** is Head of BioInformatics and Discovery Biology at EiRx Therapeutics. Current research interests include the integration of genomic and proteomic data and development of systems solutions for *in silico* modelling of genetic regulatory networks. **Ian Hayes** is the Chief Executive Officer of EiRx Therapeutics. Over a number of years, he has pioneered the application of large-scale gene-transcription analysis (genomics) to the discovery of molecular control points in disease processes, including cancer (leukaemia/lymphoma) and inflammatory disease (asthma and arteriosclerosis). His interests include identification of novel apoptosis pathways.

Carol Ward received her Ph.D. in 1998 from the Department of Medicine, University of Edinburgh. She is currently undertaking postdoctoral research work in the Centre for Inflammation Research, University of Edinburgh, investigating mechanisms underlying the control of granulocyte apoptosis in the resolution of inflammation. **Adriano G. Rossi** completed his Ph.D. in 1987 at the Department of Pharmacology, University of Glasgow. After a post-doctoral fellowship at Wake Forest University, NC, he joined the National Heart and Lung Institute, Imperial College, London. He is currently a Senior Lecturer at the Centre for Inflammation Research, University of Edinburgh Medical School and throughout his career, has investigated the mechanisms that regulate inflammatory cell biology and the induction and resolution of inflammation. **Chris Haslett** is currently Head of the MRC Centre for Inflammation Research, Director of Research for the College of Medicine and Veterinary Medicine at the University of Edinburgh and Professor of Respiratory Medicine. He moved to Edinburgh in 1990 from the Royal Postgraduate Medical School at the Hammersmith Hospital, London to become Head of the Department of Respiratory Medicine. He later became the Head of the Division of Clinical Science and Community Health and was Associate Dean of

Research in the Faculty of Medicine. He is a Fellow of the Royal Society of Edinburgh and of the Royal College of Physicians. **Ian Dransfield** obtained his Ph.D. in 1988 from the University of Sheffield after studying monocyte functional and phenotypic heterogeneity. His postdoctoral research at the Imperial Cancer Research Fund, London, contributed to the characterization of a novel antibody that defined functional activity of these adhesion receptors. Ongoing studies, at the University of Edinburgh, of the intracellular signalling events in granulocytes, apoptotic-cell clearance by phagocytes and surface molecular changes associated with neutrophil apoptosis have revealed new ways in which inflammatory cell function can be controlled, which has profound implications for the control of inflammatory diseases.

Abbreviations

ACAMP	apoptotic-cell-associated molecular pattern
AIF	apoptosis-inducing factor
ANT	adenine nucleotide translocase
Apaf-1	apoptotic protease-activating factor 1
APL	acute promyelocytic leukaemia
ARF	alternative reading frame
BH	Bcl-2 homology
CDK	cyclin-dependent kinase
cFLIP	cellular Flice-like inhibitory protein
cIAP	cellular inhibitor of apoptosis protein
CML	chronic myelogenous leukaemia
DED	death effector domain
Diablo	direct inhibitor-of-apoptosis-protein-binding protein
DISC	death-inducing signalling complex
DR	death receptor
EAE	experimental allergic encephalomyelitis
EDAR	ectodermal dysplasia receptor
EGF	epidermal growth factor
EGFR	epidermal growth factor receptor
ER	endoplasmic reticulum
FADD	Fas-associated death domain
FcR	Fc receptor
FDA	U.S. Food and Drug Administration
FLIP	Flice-like inhibitory protein
FMK	fluoromethylketone
GFP	green fluorescent protein
HDAC	histone deactylase
HSP	heat-shock protein
IκB	inhibitory κB
IAP	inhibitor of apoptosis protein
ICAM-3	intercellular cell-adhesion molecule 3
ICE	interleukin-1β-converting enzyme
IKK	inhibitory κB kinase
IL	interleukin
JNK	c-Jun N-terminal kinase
MAPK	mitogen-activated protein kinase
MEF	mouse embryo fibroblast

NF-κB	nuclear factor κB
p75-NGFR	p75-nerve growth factor receptor
PARP	polyADP-ribose polymerase
PI 3-kinase	phosphoinositide 3-kinase
PLAD	pre-ligand-binding assembly domain
PS	phosphatidylserine
PSR	PS receptor
PTP	permeability transition pore
RAR	retinoic acid receptor
RIP	receptor-interacting protein
RXR	retinoid X receptor
Smac	second mitochondrial activator of caspases
SV40	simian virus 40
TCR	T-cell receptor
TNFα	tumour necrosis factor α
TNF-R	tumour necrosis factor receptor
TRADD	tumour necrosis factor receptor 1-associated death domain protein
TRAF	tumour necrosis factor receptor-associated factor
TRAIL	tumour necrosis factor-related apoptosis-inducing ligand
TRAIL-R	tumour necrosis factor-related apoptosis-inducing ligand receptor
VDAC	voltage-dependent anion channel
v-FLIP	viral Flice-like inhibitory protein

1

Historical perspectives

James F. Curtin and Thomas G. Cotter[1]

Tumour Biology Laboratory, Department of Biochemistry, Biosciences Research Institute, University College Cork, Ireland

Abstract

Apoptosis is one of the most widely studied fields in biology and accounts for over 2% of all life science publications annually. It plays a fundamental role in development, and defects in the regulation of apoptosis are directly implicated in numerous well-known diseases including cancer, neurodegenerative disorders, tissue atrophy and auto-immune diseases. However, the field of apoptosis has humble beginnings and was neglected by biologists for much of its history. This chapter reviews the history of research in apoptosis and highlights key experiments that have contributed significantly to our current understanding of apoptosis. In addition, the topics covered in later chapters are briefly introduced.

Introduction

One of the most widely researched fields in biology since the early 1990s has been the study of apoptosis, a form of programmed cell death. Apoptosis was first defined as a collection of morphological events distinct from necrosis that were observed in a diverse range of cells and tissues undergoing cell death. These included cellular events such as cell shrinkage, membrane ruffling (known as blebbing) and packaging of the dying cell into small membrane-bound vesicles called apoptotic bodies. A variety of nuclear events were also observed that include chromatin condensation and nuclear fragmentation (Figure 1) [1]. Biochemical events specific to apoptosis were discovered later, including

[1]*To whom correspondence should be addressed (e-mail t.cotter@ucc.ie).*

degradation of DNA into regular-sized fragments [2] and exposure of a lipid called phosphatidylserine on the surface of apoptotic cells [3].

These changes in morphology and biochemistry result in cell death, and in healthy adult humans around 10 million cells undergo apoptosis daily. Normal tissue homoeostasis and function depends on correct levels of apoptosis to counteract the increase in cell numbers due to cell proliferation and also to eliminate damaged or defective cells. It is a critical event during virtually every stage of development from embryogenesis right through to adulthood. Most cells retain the ability to undergo apoptosis in response to a wide range of death-inducing stimuli. Consequently, the regulation of apoptosis is tightly controlled in cells and defects in apoptosis can give rise to a number of diseases, including cancer, neurodegenerative disorders and auto-immune disease [4].

1842 to 1972: defining 'apoptosis'

Biology is littered with examples of major discoveries that drift back into obscurity because the importance of these discoveries was not fully appreciated at the time. The work done by Gregor Mendel on the inheritance of individual characteristics in the pea plant is perhaps the most widely known example. His work went unnoticed by the general scientific community for nearly 50 years even though he had independently worked out the laws of inheritance over 30 years before gene theory was developed [5]. However, no major field in biology has been rediscovered as often as the field of apoptosis, which has been independently described at least five separate times over a period of 100 years (Figure 2).

The term apoptosis was first used in 1972 to describe a form of cell death with markedly different morphology to necrosis [1]. However, the idea that cells

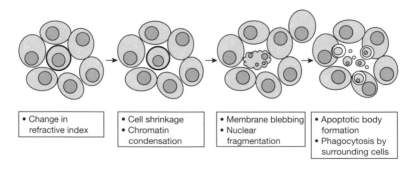

| • Change in refractive index | • Cell shrinkage • Chromatin condensation | • Membrane blebbing • Nuclear fragmentation | • Apoptotic body formation • Phagocytosis by surrounding cells |

Figure 1. Apoptosis is a form of cell death with morphology distinct from necrosis
Cells undergo a number of sequential changes in morphology when dying by apoptosis and these are outlined in this figure. The earliest identifiable change in morphology is when the refractive index of the cell changes under visible light. Cell shrinkage and condensation of the chromatin on the nuclear membrane occur next and are rapidly followed by nuclear fragmentation and plasma membrane blebbing. Apoptotic body formation is the final stage in apoptosis and the entire process is believed to take about 1 h *in vivo*.

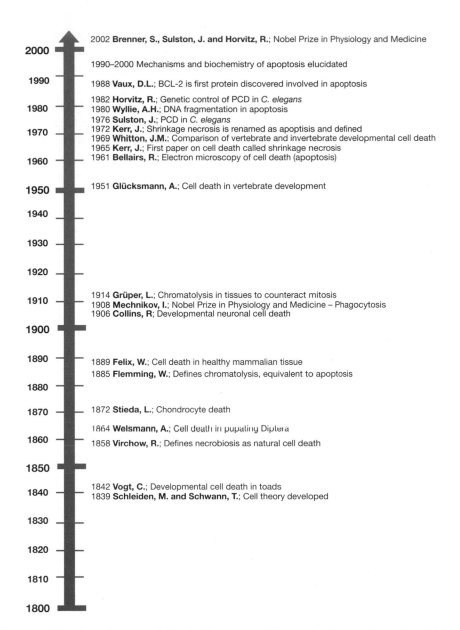

Figure 2. Timeline of apoptosis illustrating some of the major discoveries in the field of cell death

Although the concept of natural cell death was first hypothesized in 1842, the first description of apoptosis was in 1885 when Walther Flemming described a form of cell death that he called chromatolysis. This was regularly described in the literature for the next 30 years, but the field of cell death shifted in focus to phagocytosis for much of the first half of the 20th century. In the 1960s, scientists regained an interest in cell death and in 1972, apoptosis was defined for the first time by John Kerr. Interest in the field remained low until the late 1980s when it became clear that apoptosis played a key role in development and disease, and identification of Bcl-2 as an anti-apoptotic molecule marked the beginning of intensive research into the mechanisms of apoptosis throughout the 1990s. The impact that research into apoptosis had on other disciplines in biology was formally recognized when three pioneers of apoptosis research in invertebrates were awarded the Nobel Prize in Medicine and Physiology in 2002. PCD, programmed cell death.

can die naturally is a relatively old one and was first postulated 130 years earlier in 1842 by Carl Vogt after he studied development in toads. This observation was even more remarkable when one considers that Schleiden and Schwann established cell theory, which postulates that all living matter is composed of individual units or cells, only 3 years earlier in 1839. This makes the concept of natural cell death one of the oldest theories in cell biology [6]. In 1858 Rudolph Virchow described a natural form of cell death that was distinct from necrosis and called it necrobiosis [7]. The first clear morphological description of apoptosis appeared in 1885 by Walther Flemming. Drawings of cells in regressing ovarian follicles portrayed shrinkage of cells, chromatin condensation, nuclear fragmentation and formation of apoptotic bodies, and Flemming named this natural cell death chromatolysis. These are all morphological hallmarks of apoptosis and chromatolysis was regularly reviewed in the literature for over 30 years [8]. Apoptosis continued to drift into and out of fashion throughout the first half of the 20th century until an influential review on cell death was published in 1951. In this, Glücksmann described karyorrhexis and karyopyknosis as distinct forms of cell death that we now know to be events associated with apoptosis. He also suggested that cell death plays an important role in embryogenesis and development of vertebrates [9]. Consequently, some pathologists regained an interest in the mechanisms of cell death, and in 1965, the Australian pathologist John F. Kerr observed an unusual form of cell death following ischaemic injury of liver tissue, where the constituents of cells appeared to remain intact within small vesicles. He used the term shrinkage necrosis to define this form of cell death and subsequent studies using electron microscopy identified the earliest morphological events as condensation of the cytoplasm and the nuclear chromatin [10]. Kerr went on sabbatical to Aberdeen where he collaborated with Andrew Wyllie and Professor Alister Currie. During this time, they coined the term apoptosis to replace the term shrinkage necrosis and defined this particular form of cell death in a paper published in 1972 that is widely acknowledged as being a major landmark in the field of biology [1].

1972 to 1982: the first decade in apoptosis research

Researchers were initially unaware that apoptosis had been identified and described it on numerous occasions throughout the second half of the 19th century and also during the first half of the 20th century. Consequently, much of the early research simply repeated observations already made during the 19th and 20th centuries. These papers demonstrated the important role of apoptosis in development and disease, but the results were often limited to descriptions of morphology. Studies based solely on morphology have severe limitations and the interest from the general biological community during this period remained quite low. There was a significant risk that the concept of apoptosis would fade into obscurity just as chromatolysis had done 70 years previously, but for a number of important developments in the field.

Two experiments in particular were vital in maintaining an interest during this crucial first decade of research. First was the observation that DNA was degraded into nucleosomal-sized fragments during radiation-induced cell death [2]. This is due to an endogenous nuclease that is only activated in response to apoptotic stimuli and was identified 17 years later in 1997 [11]. The presence of a DNA ladder suggested that specific biochemical alterations occur in cells during apoptosis. Throughout the 1970s and 1980s work on the nematode worm *Caenorhabditis elegans* yielded the second key insight into apoptosis. *C. elegans* proved invaluable in the study of developmental cell death and painstaking analysis of the lineage of every cell in the adult worm demonstrated that exactly 131 of the 1090 somatic cells generated during development underwent programmed cell death. Both the lineage of each dying cell and the time that it underwent programmed cell death were highly reproducible, and these observations highlighted the tightly regulated nature of developmental cell death in *C. elegans* [12]. Mutants of *C. elegans* were characterized and genetic screening quickly led to the discovery that more than 12 separate genes were involved in programmed cell death [13–15].

The identification of these specific biochemical and genetic events during apoptosis sowed the seeds for the subsequent explosion in our understanding of apoptosis. The field quickly moved away from stagnant studies of cell morphology into mainstream experimental science. Eventually, apoptosis would become a centrepiece in two of the most important fields in modern biochemistry, namely cell signalling and molecular biology. The identification of an oncogene called bcl-2 that appeared to function solely as an anti-apoptotic protein in cells was a fundamental discovery that heralded the beginning of a period of intensive research into the biochemical processes underlying apoptosis.

bcl-2: the opening act

bcl-2 was initially characterized as a proto-oncogene that enhanced transformation of human lymphoma cells [16] but further work demonstrated that over-expression of *bcl-2* was found to increase the survival of haematopoietic cells by reducing the sensitivity to apoptosis. This made Bcl-2 arguably the first protein identified to be involved in apoptosis [17]. Bcl-2 was also involved in a second landmark experiment in 1992 when it was shown that over-expression could also protect against developmental cell death in *C. elegans* [18]. The ramifications of this experiment were enormous. For decades, biologists had compared the similar morphology of programmed cell death in invertebrates with apoptosis in vertebrates and speculated that the two processes were similar [19]. This simple experiment was the first conclusive proof that apoptosis in vertebrates and programmed cell death in invertebrate were conserved through evolution, and finally allowed genetic discoveries made in *C. elegans* to be applied in the study of apoptosis in mammalian cells. The unification of invertebrate and vertebrate cell death models accelerated the identification and characterization of many biochemical components of apoptosis in mammalian cells.

Unravelling the apoptotic machinery

The morphology of vertebrate and invertebrate cells undergoing apoptosis was similar regardless of the initial apoptotic signal. This suggested that a common apoptotic pathway that is activated in response to many different initial stimuli exists in cells. Previous studies with *C. elegans* in the 1980s had identified about 12 different genes that were involved in apoptosis. Some mutations completely inhibited apoptosis, while others only inhibited clearance of apoptotic bodies by surrounding cells. One gene called Ced-3 was absolutely required for apoptosis and was eventually cloned in 1993. This gene was found to share sequence homology with the mammalian protease ICE (interleukin-1β-converting enzyme; caspase 1) [20]. Since then, 12 other human caspases have been described and many have been shown to play a pivotal role in apoptosis. Unlike most other proteases, caspases have very high substrate specificity. They usually only cleave target proteins at one or two conserved sites and are absolutely required for apoptosis in response to most stimuli. Initiator caspases detect apoptosis-inducing stimuli and activate effector caspases that proceed to cleave essential enzymes and structural proteins in cells in a hierarchal protease cascade (Figure 3).

The discovery that caspases are central components in apoptosis was due primarily to genetic studies in *C. elegans*. However, research with viruses identified an entirely new class of apoptosis regulators in mammalian cells that directly inhibit caspase activity and apoptosis by steric hindrance of caspase active sites. Some baculoviruses were shown to express a protein called IAP (inhibitor of apoptosis protein) that binds with caspases and inhibit caspase activity. Herpes virus was found to express another protein called v-FLIP (viral Flice-like inhibitory protein) that shares some sequence homology with caspase-8 and was found to inhibit only death-receptor-mediated apoptosis by binding with the FADD (Fas-associated death domain) in place of caspase-8. Cellular homologues of IAP and FLIP were quickly identified and these molecules are believed to play a crucial role in regulating the sensitivity of cells to apoptotic stimuli [21,22].

The mitochondrion is particularly sensitive to many different apoptosis-inducing stimuli and dysfunction of mitochondria is associated with pore formation and mitochondrial membrane depolarization in cells. This is a reversible step, and mitochondria can repolarize and continue to generate ATP during the early stages of apoptosis. A number of proteins were identified in mitochondria that were released into the cytosol during pore formation. Some of these molecules function as molecular guillotines in the cytosol and are sequestered in mitochondria away from their cytoplasmic targets in healthy cells. One such molecule called cytochrome *c* is usually involved in oxidative phosphorylation in mitochondria. However, release of cytochrome *c* into the cytosol stimulates the formation of a multimeric complex called the apopto-

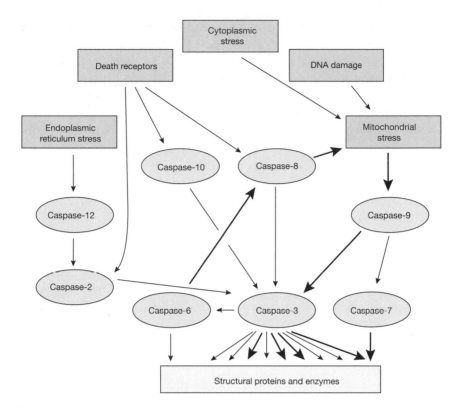

Figure 3. Caspases are central components in apoptosis
Caspases are a family of 12 aspartate proteases that have limited substrate specificity. Initiator caspases detect cellular stress signals from different sources and include caspase-2, caspase-8, caspase-9, caspase-10 and caspase-12. These initiator caspases cleave and activate effector caspases either directly or indirectly. Effector caspases, including caspase-3, caspase-6 and caspase-7, cleave a variety of downstream target proteins. This illustration highlights the central role of caspases in response to different cellular stresses.

some, where caspase-9 is recruited and can activate caspase-3 [23]. Cytochrome *c* release from mitochondria is regulated by Bcl-2 and this is believed to be the principal mode of action of the anti-apoptotic Bcl-2 family members [24]. Caspase-3, caspase-9 and Bcl-2 are homologous with Ced-3, Ced-4 and Ced-9 respectively in *C. elegans* and the conservation of this biochemical pathway helped illustrate the central role played by the mitochondrion in apoptosis.

Implicating apoptosis with disease

As the components of apoptosis were identified, it soon became clear that defects in these proteins were often associated with a number of diseases. One year after the *bcl-2* gene was cloned and characterized, a cell-surface receptor called Fas was discovered and was found to induce apoptosis and cause tumour regression when bound with anti-Fas monoclonal antibodies [25]. The gene that codes for Fas receptor was subsequently cloned and was found to share sequence homology with tumour necrosis factor receptors 1 and 2 [26]. Mutations in Fas receptor were identified in a particular strain of mice with an autosomal recessive lymphoproliferative (lpr) disorder [27]. The authors concluded correctly that these mutations were the underlying cause of auto-immune disorders in lpr mice and this was the first clear indication that apoptosis may be a key regulator of the immune system function. As a result the concept of apoptosis was introduced for the first time to many researchers studying auto-immune disease, immunosuppression, immunotolerance and AIDS.

The tumour-suppressor protein p53 had first been observed in 1979, over a decade before Fas receptor had been identified. Subsequently, p53 was found to be mutated or absent in over half of all human carcinomas [28]. While it was known that p53 could regulate transcription, the main function of p53 was not identified until researchers discovered that it induces apoptosis in response to DNA damage and defects in the cell cycle [29,30]. The prevalence of p53 mutations clearly demonstrated the importance of apoptosis during tumorigenesis. Subsequent research identified defects in apoptosis in virtually every human cancer.

Perspective: the rise and rise of apoptosis

Research in apoptosis now encompasses many diverse biological disciplines from plant science to clinical science and accounts for more than 13 000 publications annually [31]. It is remarkable that a field overlooked by the general scientific community for so long has become so important in our understanding of such a diverse range of biological processes. Early experiments identified key biochemical and genetic events and these were crucial for the subsequent explosion of research in apoptosis. Genetic studies in *C. elegans* contributed significantly to delineating the biochemical pathways in apoptosis during the 1990s. The scientific community formally acknowledged this when Sydney Brenner, John Sulston and Robert Horvitz shared the Nobel Prize in Physiology and Medicine in 2002. The Nobel Prize has arrived at an opportune time and heralds the beginning of a new era in apoptosis research. Defects in apoptosis have been attributed to many well-known incurable diseases. The most important question during the next decade of research will be how can we convert this new wealth of genetic and biochemical information into therapies that can prevent and cure these illnesses.

Summary

- *Apoptosis was first defined in 1972, but the field of cell death is much older and was originally hypothesized in 1842. Early genetic and biochemical studies of apoptosis were essential in maintaining an interest in the nascent field.*

- *The first protein implicated in apoptosis was Bcl-2, when it was found that over-expression of Bcl-2 decreased apoptosis in haematopoietic cells. Bcl-2 was also shown to inhibit apoptosis in C. elegans demonstrating that evolution had conserved the mechanisms of apoptosis between invertebrates and vertebrates.*

- *Cloning of the C. elegans gene Ced-3 allowed researchers to identify human counterparts by homology called caspases. Caspases were subsequently shown to be central components of apoptosis in mammalian cells.*

- *Two key regulators of caspases, called IAP and FLIP, were originally identified in viral DNA. Cellular homologues were quickly identified and were shown to regulate the sensitivity of cells to a number of cellular insults.*

- *Mitochondria play a pivotal role in detecting and amplifying apoptotic stimuli from nucleus, cytoplasm, endoplasmic reticulum and death receptors. Bcl-2 family members are believed to regulate apoptosis by increasing or decreasing the sensitivity of mitochondria to these apoptotic signals.*

- *The importance of apoptosis in disease was highlighted when defects in Fas receptor and p53 were shown to give rise to lymphoproliferative disorders and tumorigenesis respectively.*

- *Robert Horvitz, who made significant contributions to the field of developmental apoptosis, was awarded the Nobel Prize in Physiology and Medicine in 2002. This represents the culmination of two decades of research into the genetic and biochemical processes involved in apoptosis.*

References

1   Kerr, J.F., Wyllie, A.H. & Currie, A.R. (1972) Apoptosis: a basic biological phenomenon with wide-ranging implications in tissue kinetics. *Br. J. Cancer* **26**, 239–257

2   Wyllie, A.H. (1980) Glucocorticoid-induced thymocyte apoptosis is associated with endogenous endonuclease activation. *Nature (London)* **284**, 555–556

3   Fadok, V.A., Voelker, D.R., Campbell, P.A., Cohen, J.J., Bratton, D.L. & Henson, P.M. (1992) Exposure of phosphatidylserine on the surface of apoptotic lymphocytes triggers specific recognition and removal by macrophages. *J. Immunol.* **148**, 2207–2216

4   Reed, J.C. (2002) Apoptosis-based therapies. *Nat. Rev. Drug Disc.* **1**, 111–121

5   Nurse, P. (2000) The incredible life and times of biological cells. *Science* **289**, 1711–1716

6   Clarke, P.G. & Clarke, S. (1995) Historic apoptosis. *Nature (London)* **378**, 230

7   Gerschenson, L.E. & Geske, F.J. (2001) Virchow and apoptosis. *Am. J. Pathol.* **158**, 1543

8   Clarke, P.G. & Clarke, S. (1996) Nineteenth century research on naturally occurring cell death and related phenomena. *Anat. Embryol. (Berlin)* **193**, 81–99

9   Lockshin, R.A. & Zakeri, Z. (2001) Programmed cell death and apoptosis: origins of the theory. *Nat. Rev. Mol. Cell Biol.* **2**, 545–550

10   Kerr, J.F. (2002) History of the events leading to the formulation of the apoptosis concept. *Toxicology* **181–182**, 471–474

11   Liu, X., Zou, H., Slaughter, C. & Wang, X. (1997) DFF, a heterodimeric protein that functions downstream of caspase-3 to trigger DNA fragmentation during apoptosis. *Cell* **89**, 175–184

12   Sulston, J.E. & Horvitz, H.R. (1977) Post-embryonic cell lineages of the nematode, *Caenorhabditis elegans*. *Dev. Biol.* **56**, 110–156

13   Hedgecock, E.M., Sulston, J.E. & Thomson, J.N. (1983) Mutations affecting programmed cell deaths in the nematode *Caenorhabditis elegans*. *Science* **220**, 1277–1279

14   Ellis, H.M. & Horvitz, H.R. (1986) Genetic control of programmed cell death in the nematode *C. elegans*. *Cell* **44**, 817–829

15   Hengartner, M.O., Ellis, R.E. & Horvitz, H.R. (1992) *Caenorhabditis elegans* gene ced-9 protects cells from programmed cell death. *Nature (London)* **356**, 494–499

16   Tsujimoto, Y. & Croce, C.M. (1986) Analysis of the structure, transcripts, and protein products of bcl-2, the gene involved in human follicular lymphoma. *Proc. Natl. Acad. Sci. U.S.A.* **83**, 5214–5218

17   Vaux, D.L., Cory, S. & Adams, J.M. (1988) Bcl-2 gene promotes haemopoietic cell survival and cooperates with c-myc to immortalize pre-B cells. *Nature (London)* **335**, 440–442

18   Vaux, D.L., Weissman, I.L. & Kim, S.K. (1992) Prevention of programmed cell death in *Caenorhabditis elegans* by human bcl-2. *Science* **258**, 1955–1957

19   Whitten, J.M. (1969) Cell death during early morphogenesis: parallels between insect limb and vertebrate limb development. *Science* **163**, 1456–1457

20   Yuan, J., Shaham, S., Ledoux, S., Ellis, H.M. & Horvitz, H.R. (1993) The *C. elegans* cell death gene ced-3 encodes a protein similar to mammalian interleukin-1 beta-converting enzyme. *Cell* **75**, 641–652

21   Duckett, C.S., Nava, V.E., Gedrich, R.W., Clem, R.J., Van Dongen, J.L., Gilfillan, M.C., Shiels, H., Hardwick, J.M. & Thompson, C.B. (1996) A conserved family of cellular genes related to the baculovirus iap gene and encoding apoptosis inhibitors. *EMBO J.* **15**, 2685–2694

22   Irmler, M., Thome, M., Hahne, M., Schneider, P., Hofmann, K., Steiner, V., Bodmer, J.L., Schroter, M., Burns, K., Mattmann, C. et al. (1997) Inhibition of death receptor signals by cellular FLIP. *Nature (London)* **388**, 190–195

23   Liu, X., Kim, C.N., Yang, J., Jemmerson, R. & Wang, X. (1996) Induction of apoptotic program in cell-free extracts: requirement for dATP and cytochrome c. *Cell* **86**, 147–157

24   Kluck, R.M., Bossy-Wetzel, E., Green, D.R. & Newmeyer, D.D. (1997) The release of cytochrome c from mitochondria: a primary site for Bcl-2 regulation of apoptosis. *Science* **275**, 1132–1136

25   Trauth, B.C., Klas, C., Peters, A.M., Matzku, S., Moller, P., Falk, W., Debatin, K.M. & Krammer, P.H. (1989) Monoclonal antibody-mediated tumor regression by induction of apoptosis. *Science* **245**, 301–305

26   Itoh, N., Yonehara, S., Ishii, A., Yonehara, M., Mizushima, S., Sameshima, M., Hase, A., Seto, Y. & Nagata, S. (1991) The polypeptide encoded by the cDNA for human cell surface antigen Fas can mediate apoptosis. *Cell* **66**, 233–243

27   Watanabe-Fukunaga, R., Brannan, C.I., Copeland, N.G., Jenkins, N.A. & Nagata, S. (1992) Lymphoproliferation disorder in mice explained by defects in Fas antigen that mediates apoptosis. *Nature (London)* **356**, 314–317

28   Levine, A.J. (1997) p53, the cellular gatekeeper for growth and division. *Cell* **88**, 323–331

29   Yonish-Rouach, E., Resnitzky, D., Lotem, J., Sachs, L., Kimchi, A. & Oren, M. (1991) Wild-type p53 induces apoptosis of myeloid leukaemic cells that is inhibited by interleukin-6. *Nature (London)* **352**, 345–347

30   Lowe, S.W., Schmitt, E.M., Smith, S.W., Osborne, B.A. & Jacks, T. (1993) p53 is required for radiation-induced apoptosis in mouse thymocytes. *Nature (London)* **362**, 847–849

31   Melino, G., Knight, R.A. & Green, D.R. (2001) Publications in cell death: the golden age. *Cell Death Differ.* **8**, 1–3

2

Apoptosis and development

Justin V. McCarthy[1]

Signal Transduction Laboratory, Department of Biochemistry, Biosciences Institute, University College Cork, Cork, Ireland

Abstract

Apoptosis is an evolutionarily conserved process used by multicellular organisms to developmentally regulate cell number or to eliminate cells that are potentially detrimental to the organism. The large diversity of regulators of apoptosis in mammalian cells and their numerous interactions complicate the analysis of their individual functions, particularly in development. The remarkable conservation of apoptotic mechanisms across species has allowed the genetic pathways of apoptosis determined in lower species, such as the nematode *Caenorhabditis elegans* and the fruitfly *Drosophila melanogaster*, to act as models for understanding the biology of apoptosis in mammalian cells. Though many components of the apoptotic pathway are conserved between species, the use of additional model organisms has revealed several important differences and supports the use of model organisms in deciphering complex biological processes such as apoptosis.

Introduction

The elimination of unwanted cells by programmed cell death is an integral part of normal development and homoeostasis in multicellular organisms, including the nematode *Caenorhabditis elegans* [1,2]. It was not until the discovery of the first genetic components of this process that our appreciation for the events leading to the predetermined elimination of cells began to emerge.

[1]*E-mail jv.mccarthy@ucc.ie*

Programmed cell death, when initially described, was based on characteristic morphological events that could be readily seen in dying cells; including cell shrinkage, chromatin condensation and plasma-membrane blebbing. Later, the term apoptosis was employed to distinguish this specific type of cell death from necrosis, cell death resulting from cell injury [3].

Research in apoptosis has demonstrated that even though 'core' elements of the apoptotic machinery are widely conserved across species, animals with increased complexity possess additional proteins that control the precise execution of the apoptotic process.

Mammalian cells have two primary cell-death-inducing signalling pathways, the intrinsic 'core' and extrinsic cell-death-inducing pathways [4,5]. In the intrinsic pathway, pro-apoptotic signals lead to the release of mitochondrial proteins including cytochrome c, which binds to, and promotes the oligomerization of the adapter protein Apaf-1 (apoptotic protease-activating factor 1). Oligomerization of Apaf-1 allows the recruitment and autocatalytic activation of caspase-9 and consequently the induction of apoptosis. In the extrinsic death-inducing pathway, cell-surface death receptors of the tumour necrosis factor family induce apoptosis through homophilic interactions with intracellular adaptor proteins that mediate the autocatalytic activation of caspases [5].

Determining the physiological role of apoptosis in mammalian development requires *in vivo* studies, primarily based upon the generation and characterization of transgenic gene overexpression and gene-ablation murine models [6]. These models have contributed to our understanding of the basic apoptosis components, and how apoptosis is regulated, although they can often be complicated by genetic redundancy. With this in mind, researchers turned to lower organisms that were both developmentally characterized and genetically manipulable to provide model systems in which to study the role of apoptotic genes during development. The nematode *C. elegans* provides an ideal model for the study of cell death because of its invariant cell linage, which allows every cell to be followed during development [7,8]. Therefore, it is no surprise that *C. elegans* provided the first evidence for the genetic regulation of apoptosis [7]. The fruitfly *Drosophila melanogaster* provides a system of intermediate complexity between the nematode and mammals, sharing many mammalian components and pathways, but having less redundancy, thereby allowing easier identification of specific functions [9–11]. Perhaps the greatest advantages to studies in less complex model organisms are their powerful genetics that can be used to generate transgenic lines that either remove gene function (gene knockout) or ectopically over-express genes (transgenic animals). These systems allow the examination of genetic interactions and functional characterization of novel genes using genetic screening methods. This review will concentrate on the genetic and developmental studies that have focused on the regulation of apoptosis in *C. elegans* and *Drosophila*, which have enhanced our understanding of apoptosis in mammalian cells.

Apoptosis is an important regulator of normal development

Most studies on apoptosis have thus far focused on the identification and functional characterization of the genes and proteins involved, with little attention to its normal, though critical, role in a developing organism. Activation of apoptosis during development is regulated by many mechanisms and stimuli. Developmental studies in nematode, fruitfly and mouse models have indicated that physiological apoptotic cell death contributes to the regulation of cell number and organ size, elimination of superfluous structures, sculpting of tissues and elimination of abnormal or aged cells [1,2].

Tissue homoeostasis

The proper control of tissue and organ size involves the regulated participation of both cell division and apoptosis. For example, the development of *C. elegans* occurs through a precise, stereotypic pattern of embryonic cell divisions and cell deaths [12]. In the adult nematode, there are 959 somatic cells, but during development, an additional 131 cells undergo predetermined apoptosis. Likewise, in the development of the nervous system in vertebrates and invertebrates, excess cells are normally removed by apoptosis [13,14]. It is proposed that these excess cells may be required for the developmental stage-specific establishment of appropriate patterns of cell migration and morphogenesis, and may not be functionally required later. Perhaps the best-studied example of neuronal cell death coincides with synaptogenesis in post-mitotic neurons [14]. After neurons have differentiated and extended axons to their targets, approx. 50% of the original cells are eliminated. In this case, the number of cells that die is not predetermined, but influenced by a limiting amount of neurotrophic factor (i.e. nerve growth factor and neurotrophins) produced by target tissues. Such studies have led to the formulation of the trophic-factor hypothesis [15].

Formation and deletion of structures

Apoptosis is important for the proper formation of digits (fingers and toes) of the hand and foot where, during embryonic development, inter-digital cells are removed. A deficiency in apoptosis manifests as webbed fingers in infants [1]. Likewise, many larval tissues that are no longer required in the adult are removed by apoptosis. During the development of amphibians such as frogs, in transition to an adult, the tadpole undergoes a major structural remodelling whereby the tail, notochord and intestine are deleted [16]. Insect metamorphosis involves extensive tissue rearrangement and many larval cells either die by apoptosis or differentiate into adult cells.

Removal of abnormal cells

During development and subsequent tissue homoeostasis, the maintenance of DNA fidelity and proper cell function is of critical importance. The propagation of cells that carry mutations can lead to deleterious characteristics such as aberrant cell growth or auto-immune disease. During cell-cycle progression, the checkpoint machinery detects aberrant DNA replication and mitotic events, and if repair is not possible, the cells are later eliminated by apoptosis. In the developing immune system, any lymphocytes that produce self-reactive receptors are likewise eliminated by apoptosis [17].

Phagocytosis and removal of apoptotic cells

At any one time or developmental stage, very few apoptotic cells can be detected *in vivo*. This is due to the rapid phagocytic removal and degradation of apoptotic cells by neighbouring cells and macrophages. The engulfment process, which removes the dying cells before they can lyse and release potentially harmful cytoplasmic contents, is important for tissue remodelling. The mechanisms that regulate the phagocytosis of dying cells have been studied extensively in *C. elegans*, and functional mammalian homologues have subsequently been identified. Seven genes identified in *C. elegans* are involved in two redundant phagocytosis or engulfment pathways [8,18]. Mutations in any single gene within a pathway do not alter phagocytosis, but a mutation in genes in both pathways results in defective engulfment.

Global regulators of apoptosis in *C. elegans*

C. elegans has proved to be very useful for investigating how patterns of cell fate are established during animal development because it has a completely defined and largely invariant cell lineage. That is, the division, differentiation and development fate of each individual cell follows a precise and predetermined pathway of development. Individual cells can be observed easily in live animals, comparative animal-to-animal studies are readily performed and *C. elegans* is well suited for genetic and molecular analysis. During the development of an adult *C. elegans*, 131 out of the total 1090 cells undergo apoptotic cell death in a lineage-specific manner. Genetic studies in *C. elegans* have defined a variety of single-gene mutations that have specific effects on apoptosis. Analysis of the genes defined by these mutations have revealed that apoptosis is an active process that requires the function of specific genes; some are required to cause cell death, while others protect cells from dying. The genetic dissection of this developmental cell death in *C. elegans* has led to the identification of three groups of genes involved in this process. The first group of genes includes *ces-1* and *ces-2* (*ces*, cell-death specification) and affects the death of specific types of cells. The second group of genes affects most, if not all, of the 131 cells undergoing apoptosis, and is therefore defined as the core apoptosis pathway. These highly conserved regulators, which include four genes, *egl-1* (*egl*, egg-laying-defective), *ced-3* (*ced*, cell-death abnormal), *ced-4* and *ced-9*, are central to developmental apoptosis in *C. elegans* [7,19]. The last group of genes, which includes *nuc-1* (*nuc*, nuclease) *ced-1*, *ced-2*, *ced-5*, *ced-6*, ced-7, ced-10 and ced-12, is involved in the degradation of DNA and phagacytosis of apoptotic cells. Among the three groups of cell-death genes, those involved in the execution phase of apoptosis have been most extensively studied, with biochemical studies suggesting that expression of egl-1, ced-3 and ced-4 are required for the induction of apoptosis, whereas expression of ced-9 is necessary to inhibit cell death [8].

Conservation of apoptotic proteins and their regulators

Homologues of the nematode core apoptotic pathway genes are conserved throughout evolution, along with additional activators, effectors and inhibitors of cell death (Table 1). Gene-interaction studies have defined a genetic pathway and have ordered the functions of egl-1, ced-9, ced-3 and ced-4 (Figure 1). The ced-3 gene encodes a defining member of a continuously growing family of specific cystine proteases, termed caspases (cysteinyl aspartate-specific proteases). The first identified mammalian homologue of ced-3 was ICE (interleukin-1β-converting enzyme), responsible for the processing of pro-interleukin-1β to the active cytokine. Subsequent studies have lead to the identification of a total of 14 mammalian caspases (caspases 1–14) [20]. Caspases are broadly grouped into two classes based on their proteolytic specificities and function in either cytokine maturation or apoptosis. Apoptotic caspases are further subdivided into two groups, the initiator or long-pro-domain caspases and the effectors or short-pro-domain caspases. The initiator caspases are involved at the instigation of a proteolytic caspase cascade and amplification of death signals that result in the activation of the effector caspases. In Drosophila, seven caspases have been identified thus far, Dcp-1, Dredd/Dcp-2, Drice, Dronc, Decay, Strica/Dream and Damm/Daydream [11].

The nematode ced-4 gene functions in the activation of ced-3 and only one mammalian homologue, Apaf-1, has been characterized to date, which, like ced-4, functions in the activation of caspases (Figure 1). In C. elegans, the CED-4 protein is normally localized to the mitochondria through an interaction with CED-9, unless EGL-1 is expressed. If EGL-1 is expressed, the interaction between CED-4 and CED- 9 is disrupted, and CED-4 translocates to the nuclear membrane where it activates CED-3, which subsequently induces apoptosis. Both CED-4 and APAF-1 require dATP for caspase activation, but mammalian Apaf-1 also requires mitochondrial cytochrome c [21]. Drosophila has recently been shown to have a ced-4/Apaf-1 homologue, Ark (previously called dapaf-1, Dark, hac-1). Importantly, similar to APAF-1 and CED-4, loss-of-function Ark mutants have reduced developmental apoptosis.

The nematode CED-9 and EGL-1 proteins belong to a growing family of proteins with homology with the mammalian anti-apoptotic protein Bcl-2 [4,22]. This family is functionally divided into pro- and anti-apoptotic subgroups. CED-9 is homologous with the anti-apoptotic subgroup, whereas

Table 1. Conservation of sequence or function in apoptosis proteins and regulators across species

IAP, inhibitor of apoptosis protein; cIAP, cellular IAP; Diablo, direct IAP-binding protein; Smac, second mitochondrial activator of caspases; AIF, apoptosis-inducing factor; RGH, reaper-grim-hid; CSP, caspase homologue; HID, head involution defective; XIAP, X-linked inhibitor of apoptosis; NAIP, neuronal apoptosis inhibitor protein; CAD, caspase-activated deoxyribonuclease; ICAD, inhibitor of CAD; DFF, DNA fragmentation factor; SREC, scavenger receptor expressed by endothelial cells; ABC1, ATP-binding cassette transporter 1 gene; CRK, CT10-regulator of kinase; DOCK, dedicator of cytokinesis; ELMO, engulfment and cell motility; WAH, C. elegans homologue of human AIF protein.

Family	C. elegans	Drosophila	Mammals
Transcription	Tra-1, Ces-1, Ces-2		
Caspase	CED-3, CSP-1, CSP-2	Dcp-1, Dredd/Dcp-2, Drice, Dronc, Decay, Strica/Dream, Damm/Daydream	Caspases 1–14
Bcl-2	CED-9, EGL-1	Debcl-1/Drob-1/Dborg-1/Dbok, Buffy/Dborg-2	Bcl-2, Bcl-x, Bcl-w, Mcl-1, Diva, Bax, Bak, Bok, Bik, Bid, Bad, Hrk/DP5, Bim, Bnip, Nix, A1
Apaf-1	CED-4	Ark/Dark/Hac-1/dApaf-1	Apaf-1
IAP	Bir-1, Bir-2	dIAP-1, dIAP-2, dBruce, Deterin	cIAP1, cIAP2, XIAP, NAIP, Survivin, Bruce, ml-IAP, c-AP2
RGH domain		Reaper, Grim, HID, Sickle	Smac/Diablo, Omi/HtrA2
Endonuclease	Nuc-1, CPS-6		ICAD/DFF40, CAD/DFF45,Endo-G
Oxidoreductase	WAH-1		AIF
Engulfment	CED-1, CED-6, CED-7, CED-2, CED-5, CED-10, CED-12		SREC, ABC1, hCED-6, CRKII, DOCK180, RAC1, ELMO

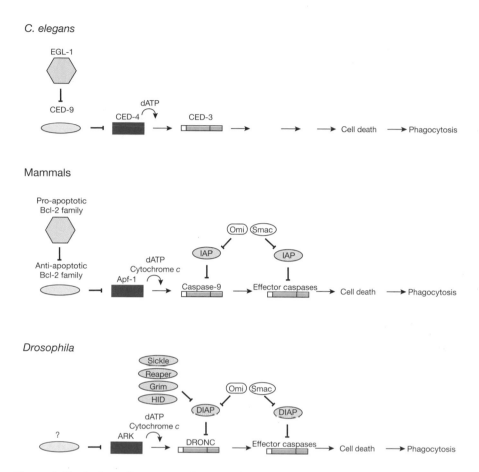

Figure 1. Evolutionarily conserved apoptotic pathways

The mammalian proteins that participate in the intrinsic apoptosis pathway and homologus proteins in *C. elegans* and *Drosophila* are represented. In *C. elegans*, the initiator caspase CED-3 is activated by CED-4, but is inhibited by CED-9. In mammals, the intrinsic pathway involves the translocation to mitochondria of pro-apoptotic Bcl-2 family members, which result in the release of mitochondrial cytochrome *c*, oligomerization of Apaf-1, activation of caspase-9 and subsequent activation of effector caspases. The IAPs bind to activated caspases, and thereby modulate the induction of apoptosis. The mitochondrial proteins Omi and Smac/Diablo also interact with the IAPs and prevent them from inhibiting caspase activation, thereby inducing apoptosis. DIAP, *Drosophila* IAP; DRONC, *Drosophila* Nedd2-like caspase.

EGL-1 has mammalian counterparts in the pro-apoptotic BH (Bcl-2 homology) 3-domain-only protein group. In addition to regulating activation of CED-4/APAF-1, family members have been shown to play a pivotal role in the regulation of mitochondrial homoeostasis and release of pro-apoptotic factors, such as AIF (apoptosis-inducing factor), Smac/Diablo [second mitochondrial activator of caspases/direct IAP (inhibitor of apoptosis protein)-binding protein], Omi/HtrA2, endonuclease G and cytochrome *c*, which mediate various aspects of apoptosis in mammalian cells [4]. These proteins have been shown to function through caspase-dependent pathways (Smac/Diablo and cytochrome *c*), caspase-independent pathways (AIF and endonuclease G) or both (Imi/HtrA2). The nematode homologues of endonuclease G and AIF, CSP-6 (caspase homologue 6) and WAH-1 (*C. elegans* homologue of human AIF) respectively, promote apoptotic DNA degradation and represent some of the mitochondrial proteins implicated in invertebrate apoptosis [23]. Thus endonuclease G/CSP-6 and AIF/WAH-1 provide another example of a single conserved apoptotic signalling pathway between vertebrates and invertebrates. Following the lead of studies in *C. elegans*, similar genetic studies in *Drosophila* have also validated the conservation of apoptosis signalling events through the identification of several additional regulators. A region of the *Drosophila* genome that contains four genes, *sickle*, *reaper*, *grim* and *head involution defective* (*hid*) has been shown to be essential for virtually all apoptosis during *Drosophila* embryogenesis and development [11,24,25]. These genes have minimal sequence homology with any known mammalian proteins, apart from a small region of 15 amino acids shared by all four genes. Evidence for the existence of vertebrate homologues of these genes comes from the observations that these proteins are able to functionally interact with a family of inhibitors of apoptosis, the IAPs [26]. The IAPs were first identified in baculovirus, but have since been shown to be important regulators of apoptosis in *Drosophila* and mammals (Figure 1) [26].

Though it has no sequence homology with the *Drosophila* genes *reaper*, *grim* or *hid*, one mammalian protein, Smac/Diablo, does share many functional similarities. Like Reaper, Grim and Hid, Smac/Diablo also interacts with the IAPs and prevents them from inhibiting caspase activation, thereby inducing apoptosis (Figure 2). Furthermore, Smac/Diablo interacts with the BIR (baculovirus IAP repeat) domain of the IAPs, similar to Reaper, Grim and Hid.

Developmental regulation of cell death

Although genetic and biochemical studies have considerably advanced our understanding of the genes involved in the execution and regulation of apoptosis, little attention has been given to the regulation of apoptosis in developing animals. One perplexing question is how do cells decide to undergo apoptosis? In mammals, cell interactions act to initiate some forms of programmed cell death, while in *C. elegans*, very few cell deaths depend upon cell interactions, with many being cell-autonomous [27]. There are two ways in which autonomous cell death can be initiated. First, apoptosis might be triggered by an underlying physiological cellular defect, such as DNA damage or a defect in differentiation. Secondly, developmental apoptosis might be a differentiation fate expressed as a consequence of the normal process of terminal differentiation. In this way, developmental apoptosis is similar in a number of ways to adopting a state of differentiation, such as becoming a neuron or muscle cell. Like other developmental fates in *C. elegans*, cell death occurs throughout the cell lineage, with the majority of cell deaths occurring

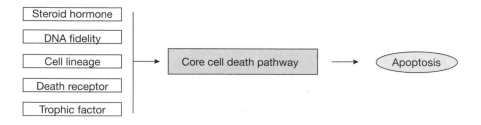

Figure 2. Pathways leading to apoptotic cell death
Apoptosis is activated by many stimuli, including steroid hormones, DNA damage, cell-lineage information, membrane-bound death receptors and extracellular survival factors. These signals converge into the core cell-death machinery, activating caspases, cell degradation and phagocytosis.

during the developmental period when most other cells terminally differentiate. Also similar to other cell fates in *C. elegans*, developmental apoptosis is observed in an invariant pattern with the same linearly equivalent cells undergoing apoptosis from animal to animal. This reproducibility in the cells that die suggests that developmental apoptosis is not a result of physiological cellular defects, but rather indicates that developmental apoptosis is no different from developmental cell fates in general. The integration of many cellular signals, including extracellular survival factors [15,17], cell-surface death receptors [5,17], cell-lineage determinants [8], steroid hormones [10,28] and intracellular and extracellular stress signals [29], determines when apoptosis occurs during development (Figure 2).

Studies have recently provided abundant information on how these signalling events initiate the regulated removal of specific cells by apoptosis.

Extracellular survival factors

Withdrawal of growth-factor signals has long been recognized as an important mediator of cell death and has led to the trophic-factor hypothesis [15]. During development in vertebrate nervous systems, excessive numbers of glia and postsynaptic neurons are produced that ultimately die by apoptosis [14,15]. The number of cells that die is not predetermined, but is influenced by a limiting amount of neurotrophins or extracellular stimuli produced by target tissues. During vertebrate synaptogenesis in post-mitotic neurons, following neuronal differentiation and axonal extension to target tissues, approx. 50% of the original cells are eliminated. Recently, the mechanism underlying trophic factor signalling was determined using *Drosophila* development as a model system [30].

During the development of *Drosophila*, approx. 10 midline glial cells are generated in each embryonic segment and function to separate axon tracts. Most of these glial cells die at a specific embryonic stage, dictated by the loss of SPITZ, a neuronal secreted ligand of the EGFR (epidermal growth factor receptor). In the presence of SPITZ, the EGFR initiates a Ras/MAPK (mitogen-activated protein kinase) anti-apoptotic signalling pathway, which suppresses expression of the pro-apoptotic protein HID [31]. Therefore, loss of

SPITZ signal enables HID (head involution defective) to activate apoptosis by interacting with DIAP1 (*Drosophila* IAP1) and releasing its inhibition of caspases [30]. Whether equivalent regulation by survival signals occurs in vertebrates remains to be determined.

Cell-lineage information

There are two forms of the nematode *C. elegans*, hermaphrodites and males. During development of the male, expression of the sex-determining gene *tra-1*, which encodes a zinc-finger transcription regulator, regulates apoptosis of hermaphrodite-specific neurons. The Tra-1 protein represses transcription of the pro-apoptotic *egl-1* gene and thereby prevents induction of apoptosis by EGL-1 in hermaphrodites [32]. During male development EGL-1 induces apoptosis of hermaphrodite-specific neurons by sequestering the anti-apoptotic protein CED-9. EGL-1 has significant sequence homology with other members of the BH3 subfamily of pro-apoptotic Bcl-2 family members, suggesting that similar mechanisms may be utilized in other species.

Steroid hormones

During *Drosophila* embryogenesis and development, increasing numbers of apoptotic cells are observed throughout the embryo, particularly in the developing nervous system [10,32]. Later, during metamorphosis, abundant tissue and structural reorganization involves extensive apoptosis that is regulated by specific changes in the concentration of the steroid hormones [11,28]. The steroid hormone ecdysone regulates both cell differentiation and cell death during insect metamorphosis, by hierarchical transcriptional regulation of a number of genes, including the zinc-finger family of transcription factors. These genes in turn regulate the transcription of a number of downstream genes. DRONC (*Drosophila* Nedd2-like caspase), a key apical caspase in *Drosophila*, is the only known caspase that is transcriptionally regulated by ecdysone during development [33,34]. Decreased ecdysone concentrations also induce neuronal cell death by altering the transcription of the pro-apoptotic genes *reaper*, *grim* and *hid* [32]. Similarly, in mammals, withdrawal of androgens induces apoptosis in the prostate gland, and increased production of the male sex hormone testosterone induces apoptosis in mammary cells in males [1]. Further increases in the *Drosophila* steroid hormone ecdysone at the late larval/early pupal stage also induces cell death of larval tissues including the midgut and salivary glands, which are not needed in the adult [35].

Developmental expression and modification of apoptotic components

As outlined above, apoptosis is a tightly regulated process where aberrant regulation leads to gross developmental defects or disease. It is important to maintain low levels of pro-apoptotic proteins in the cytoplasm under normal growth conditions, but to rapidly induce their expression or activity when cells are programmed or stimulated to undergo apoptosis. For this reason, the function of many genes and proteins involved in apoptosis are developmentally controlled both at the transcriptional level and via several post-translational mechanisms. For example, at the protein level, Apaf-1 may be modulated by proteolytic cleavage, subcellular localization and by association with protein modifiers. Transcriptional regulation of *Apaf-1* has been implicated during development of the mammalian central nervous system. Members of the bcl-2/ced-9/egl-1 family are also modulated by transcriptional and post-translational mechanisms [29]. In *C. elegans*, expression of *egl-1* in hermaphrodite-specific neurons is normally repressed by the transcription factor TRA-1 in hermaphrodites, but not in males [32]. In mammals, at least four Bcl-2 family members are subject to transcriptional control [29]. During embryogenesis, Hrk/DP5 expression is up-regulated upon withdrawal of nerve growth factor or treatment with amyloid β protein, and its levels peak at the time when these cells are committed to die. Phosphorylation-induced changes in protein conformation, which cause release from an inactive complex, increase affinity for the formation of homo- and hetero-dimers, and is central to the regulation of members of the bcl-2/ced-9/egl-1 family [29]. *Drosophila hid* is negatively regulated at the transcriptional and post-transcriptional levels by the EGFR/Ras/Raf/MAPK pathway [30,31]. By conducting genetic interaction studies in transgenic models of *Drosophila*, it was also demonstrated that the EGFR pathway promotes cell survival by transcriptional repression of *hid* and that the MAPK phosphorylation sites in Hid are critical for this response [30]. Additional *in vitro* promoter analysis studies of *reaper*, *grim* and *hid* genes revealed that steroid hormones, developmental signals and growth factors modulate expression of these genes [9–11,28].

In vivo gene-manipulation studies

Developmental genetics in model systems, including *C. elegans* and *Drosophila*, have helped to identify and order the components of apoptotic pathways. An even more complex network of apoptotic pathways has evolved in higher organisms that possess multiple homologues within each set of cell-death regulators. Whereas biochemical studies provide details of molecular mechanisms of apoptosis, genetic models reveal the essential physiologic roles. Comparison of transgenic or gene-knockout mice with their wild-type littermates has helped to elucidate mammalian apoptotic pathways and identify the principal effect of each cell-death regulator, thereby enabling an assessment of the role played by specific genes in apoptosis during mammalian development [6]. For example, Apaf-1-null animals suffer a variety of developmental defects and either die *in utero* or shortly after birth [36]. *Apaf-*

1-null embryos exhibit severe cranio-facial deformation, neural and brain overgrowth and retain inter-digital webs; all features of insufficient or compromised apoptotic cell death [36]. These studies demonstrate that *Apaf-1* is critical for apoptosis in the central nervous system and for normal brain development, although the role of *Apaf-1* in other forms of developmental apoptosis is still unclear. Unlike mice, *Drosophila* mutants that lack *Apaf-1* expression are viable, indicating the need for some caution when comparing developmental apoptotic processes between species, particularly between *Drosophila* and humans.

Caspase-8-null mice die *in utero* as a result of defective development of heart muscle and display fewer haematopoietic progenitor cells, suggesting that the FADD (Fas-associated death domain)/caspase-8 pathway is absolutely required for growth and development of specific cell types. Similarly, animals with mutations in *caspase-3* and *caspase-9* die early in development and show defects in apoptosis in the nervous system, resulting in brain overgrowth [6]. From these studies, it is evident that cell death is an important component of development in the nervous system, brain and cardiovascular system, but more detailed studies are required to understand the precise role that dying cells have in the formation of these structures and systems.

Conclusions

Normal development is tightly regulated by cell division and apoptotic cell death. Genetic and biochemical studies have provided detailed information on the proteins and enzymes responsible for the activation, regulation and mechanisms of apoptosis. Because of the close association between aberrant cell death and diseases (cancer, auto-immune disease), the therapeutic modulation of apoptosis has become an area of intense research, but with this comes the demand for a more thorough understanding of apoptosis in whole organisms. In recent years, researchers have begun to look at the regulation of apoptosis in more integrated systems and model organisms. This approach was successfully demonstrated for *C. elegans* where the core apoptotic pathway was first identified and characterized, and has now been applied to *Drosophila* and other model organisms. Collectively, these studies have identified and characterized the basic genes and proteins involved in mammalian apoptosis, and demonstrate an impressive universality of core apoptotic proteins, mechanisms and regulation among developmentally divergent species. Future studies of developmental apoptosis will determine more similarities and differences between the mechanisms that mediate cell death under physiological conditions, thereby providing invaluable information in the diagnosis of disease and design of appropriate therapeutics.

Summary

• *Physiological apoptotic cell death is an evolutionarily conserved gene-directed process essential for normal development, contributing to the regulation of cell numbers and organ size, formation or elimination of superfluous structures, sculpting of tissues, and elimination of abnormal or aged cells.*

• *Mammalian cells have two primary apoptotic death-inducing signalling pathways, the intrinsic 'core' and extrinsic cell death-inducing pathways.*

• *Several factors participate in the developmental activation of apoptosis, including extracellular survival factors, cell-surface death receptors, cell-lineage information, steroid hormones, and responses to intracellular and extracellular stress signals.*

• *Determining the physiological role of apoptosis in mammalian development requires in vivo studies, primarily attempted by the generation and characterization of transgenic and gene-ablation murine models.*

• *Even though 'core' elements of the apoptotic machinery are widely conserved across species, animals with increased complexity possess an array of divergent regulatory events that control apoptosis, indicating the need for some caution when comparing developmental apoptotic processes between species.*

References

1. Baehrecke, E.H. (2002) How death shapes life during development. *Nat. Rev. Mol. Cell Biol.* **3**, 779–787

2. Vaux, D.L. & Korsmeyer, S.J. (1999) Cell death In development. *Cell* **96**, 245–254

3. Kerr, J.F., Wyllie, A.H. & Currie, A.R. (1972) Apoptosis: a basic biological phenomenon with wide-ranging implications in tissue kinetics. *Br. J. Cancer* **26**, 239–257

4. Zimmermann, K.C., Bonzon, C. & Green, D.R. (2001) The machinery of programmed cell death. *Pharmacol. Ther.* **92**, 57–70

5. Wajant, H. (2002) The Fas signaling pathway: more than a paradigm. *Science* **296**, 1635–1636

6. Ranger, A.M., Malynn, B.A. & Korsmeyer, S.J. (2001) Mouse models of cell death. *Nat. Genet.* **28**, 113–118

7. Ellis, H.M. & Horvitz, H.R. (1986) Genetic control of programmed cell death in the nematode *C. elegans. Cell* **44**, 817–829

8. Liu, Q.A. & Hengartner, M.O. (1999) The molecular mechanism of programmed cell death in *C. elegans. Ann. N.Y. Acad. Sci.* **887**, 92–104

9. Muqit, M.M. & Feany, M.B. (2002) Modelling neurodegenerative diseases in *Drosophila*: a fruitful approach? *Nat. Rev. Neurosci.* **3**, 237–243

10. Abrams, J.M. (1999) An emerging blueprint for apoptosis in *Drosophila. Trends Cell Biol.* **9**, 435–440

11. Richardson, H. & Kumar, S. (2002) Death to flies: *Drosophila* as a model system to study programmed cell death. *J. Immunol. Methods* **265**, 21–38

12. Ellis, R.E., Yuan, J.Y. & Horvitz, H.R. (1991) Mechanisms and functions of cell death. *Annu. Rev. Cell Biol.* **7**, 663–698

13. Horvitz, H.R., Sternberg, P.W., Greenwald, I.S., Fixsen, W. & Ellis, H.M. (1983) Mutations that affect neural cell lineages and cell fates during the development of the nematode *Caenorhabditis elegans. Cold Spring Harb. Symp. Quant. Biol.* **48**, 453–463

14. Nijhawan, D., Honarpour, N. & Wang, X. (2000) Apoptosis in neural development and disease. *Annu. Rev. Neurosci.* **23**, 73–87

15. Barde, Y.A. (1989) Trophic factors and neuronal survival. *Neuron* **2**, 1525–1534

16. Shi, Y.B. & Ishizuya-Oka, A. (1996) Biphasic intestinal development in amphibians: embryogenesis and remodeling during metamorphosis. *Curr. Topics Dev. Biol.* **32**, 205–235

17. Krammer, P.H. (2000) CD95's deadly mission in the immune system. *Nature (London)* **407**, 789–795

18. Conradt, B. (2001) Cell engulfment, no sooner ced than done. *Dev. Cell* **1**, 445–447

19. Conradt, B. & Horvitz, H.R. (1998) The *C. elegans* protein EGL-1 is required for programmed cell death and interacts with the Bcl-2-like protein CED-9. *Cell* **93**, 519–529

20. Shi, Y. (2002) Mechanisms of caspase activation and inhibition during apoptosis. *Mol. Cell* **9**, 459–470

21. Li, P., Nijhawan, D., Budihardjo, I., Srinivasula, S.M., Ahmad, M., Alnemri, E.S. & Wang, X. (1997) Cytochrome c and dATP-dependent formation of Apaf-1/caspase-9 complex initiates an apoptotic protease cascade. *Cell* **91**, 479–489

22. Bouillet, P. & Strasser, A. (2002) BH3-only proteins – evolutionarily conserved proapoptotic Bcl-2 family members essential for initiating programmed cell death. *J. Cell Sci.* **115**, 1567–1574

23. Wang, X., Yang, C., Chai, J., Shi, Y. & Xue, D. (2002) Mechanisms of AIF-mediated apoptotic DNA degradation in *Caenorhabditis elegans*. *Science* **298**, 1587–1592

24. Srinivasula, S.M., Datta, P., Kobayashi, M., Wu, J.W., Fujioka, M., Hegde, R., Zhang, Z., Mukattash, R., Fernandes-Alnemri, T., Shi, Y. et al. (2002) sickle, a novel *Drosophila* death gene in the reaper/hid/grim region, encodes an IAP-inhibitory protein. *Curr. Biol.* **12**, 125–130

25. Wing, J.P., Karres, J.S., Ogdahl, J.L., Zhou, L., Schwartz, L.M. & Nambu, J.R. (2002) *Drosophila* sickle is a novel grim-reaper cell death activator. *Curr. Biol.* **12**, 131–135

26. Salvesen, G.S. & Duckett, C.S. (2002) IAP proteins: blocking the road to death's door. *Nat. Rev. Mol. Cell Biol.* **3**, 401–410

27. Sulston, J.E. & White, J.G. (1980) Regulation and cell autonomy during postembryonic development of *Caenorhabditis elegans*. *Dev. Biol.* **78**, 577–597

28. Jiang, C., Baehrecke, E.H. & Thummel, C.S. (1997) Steroid regulated programmed cell death during *Drosophila* metamorphosis. *Development* **124**, 4673–4683

29. Puthalakath, H. & Strasser, A. (2002) Keeping killers on a tight leash: transcriptional and post-translational control of the pro-apoptotic activity of BH3-only proteins. *Cell Death Differ.* **9**, 505–512

30. Bergmann, A., Tugentman, M., Shilo, B.Z. & Steller, H. (2002) Regulation of cell number by MAPK-dependent control of apoptosis: a mechanism for trophic survival signaling. *Dev. Cell* **2**, 159–170

31. Bergmann, A., Agapite, J., McCall, K. & Steller, H. (1998) The *Drosophila* gene hid is a direct molecular target of Ras-dependent survival signaling. *Cell* **95**, 331–341

32. Conradt, B. & Horvitz, H.R. (1999) The TRA-1A sex determination protein of *C. elegans* regulates sexually dimorphic cell deaths by repressing the egl-1 cell death activator gene. *Cell* **98**, 317–327

33. Cakouros, D., Daish, T., Martin, D., Baehrecke, E.H. & Kumar, S. (2002) Ecdysone-induced expression of the caspase DRONC during hormone-dependent programmed cell death in *Drosophila* is regulated by Broad-Complex. *J. Cell Biol.* **157**, 985–995

34. Dorstyn, L., Colussi, P.A., Quinn, L.M., Richardson, H. & Kumar, S. (1999) DRONC, an ecdysone-inducible *Drosophila* caspase. *Proc. Natl. Acad. Sci. U.S.A.* **96**, 4307–4312

35. Baehrecke, E.H. (2000) Steroid regulation of programmed cell death during *Drosophila* development. *Cell Death Differ.* **7**, 1057–1062

36. Yoshida, H., Kong, Y.Y., Yoshida, R., Elia, A.J., Hakem, A., Hakem, R., Penninger, J.M. & Mak, T.W. (1998) Apaf1 is required for mitochondrial pathways of apoptosis and brain development. *Cell* **94**, 739–750

3

Caspases: the enzymes of death

Boris Zhivotovsky[1]

Institute of Environmental Medicine, Division of Toxicology, Karolinska Institutet, Box 210, SE-171 77 Stockholm, Sweden

Abstract

The caspases are a unique family of cysteine proteases, which cleave proteins next to an aspartate residue. Among all known mammalian proteases, only the serine protease granzyme B has similar substrate specificity. In addition to a central role of caspases in the initiation and execution phases of apoptosis, these enzymes have some other non-apoptotic functions in living cells. During apoptosis, upon activation, caspases cleave specific substrates and thereby mediate many of the typical biochemical and morphological changes in apoptotic cells, such as cell shrinkage, chromatin condensation, DNA fragmentation and plasma-membrane blebbing. Thus, detection of activated caspases can be used as a biochemical marker for apoptosis induced by diverse stimuli in many types of cells.

Introduction

At the end of the 1980s a novel mammalian protease, ICE (interleukin-1β-converting enzyme; referred to as caspase-1), was identified and has been shown to play an important role in inflammation. Several years later Robert Horvitz's group reported that ICE is related to a *Caenorhabditis elegans* death gene product, CED-3 [1]. This seminal observation suggested that the

[1]*E-mail Boris.Zhivotovsky@imm.ki.se*

molecular mechanisms of cell death might be highly conserved and that cysteine proteases might be essential for the regulation of the cell-death process. Since 1993 a family of related proteases has been described. This family, termed the caspases (for cysteinyl aspartate-specific proteases), indeed has been shown to play a central role in initiation and execution of apoptosis [2]. The importance of caspases for the apoptotic process was documented by several findings; (i) over-expression of caspases efficiently kills cells; (ii) synthetic or natural inhibitors of caspases effectively inhibit apoptosis induced by diverse stimuli; and (iii) knock-out animals lacking certain caspases show profound defects in apoptosis. Given the important role of this family in cell death, it is not surprising to find that these proteins might have some other functions. Indeed, non-apoptotic functions of caspases have also been described. Thus caspases are involved in the processing of cytokines during inflammation, in proliferation of T-lymphocytes as well as in terminal differentiation of lens epithelial cells and keratinocytes (for a review see [3]). It is important to note that the role of caspases in normal cellular processes as well as in apoptosis is complex in that the caspases may have redundant functions. Moreover, the extent and kinetics of caspase activation are dependent on both the apoptotic stimuli and the cell type.

Structural and functional organization of caspases

The caspase family of proteases consists of at least 14 mammalian members. Except caspase-11 (mouse), caspase-12 (mouse) and caspase-13 (bovine), all other enzymes are of human origin (Figure 1). A phylogenic analysis revealed that the gene family is composed of two major subfamilies, which are related to either caspase-1 (ICE) or the mammalian homologues of CED-3 [4,5]. Although over-expression of almost all caspases kills cells, the majority of proteins belonging to the caspase-1 subfamily are involved in processing of cytokines. In contrast, proteins from CED-3 subfamily appear to be involved primarily in cell death. Based on sequence similarities, CED-3 subfamily can be further subdivided into two groups, namely the caspase-3 and caspase-2 subfamilies. Caspases are constitutively expressed in the majority of investigated cell types as inactive pro-enzymes (zymogens) that become proteolytically processed and activated in response to a variety of pro-apoptotic stimuli (Figure 2). The pro-caspases (32–56 kDa) contain several domains: an N-terminal pro-domain, a large subunit (17–21 kDa) and a small subunit (10–13 kDa); some proteins also have a short linker region between the large and small subunits [6]. Caspase activation involves proteolytic processing of the pro-enzyme at a specific aspartic acid residue site between the large and small subunits, and in many cases, pro-domains are proteolytically removed. The active site is formed by a heterodimer containing one large and one small subunit, and the fully active caspase is a tetramer composed of two such heterodimers (Figure 2).

Based on the size of pro-domains, pro-caspases can also be subdivided into two groups: those with short pro-domains (less than 30 amino acids) and those with long pro-domains (more than 100 amino acids). Pro-caspases 3, 6 and 7 contain short pro-domains and the remaining caspases have long pro-domains. All caspases with short pro-domains fulfil the role of effector enzymes. Although pro-caspase-14 also belongs to the short-pro-domain caspases, this enzyme does not participate directly in the apoptotic cascade. Its activity has been implicated in the specific form of programmed cell death in the skin [7]; however, the molecular mechanism of pro-caspase-14 maturation is unknown. The rest of the enzymes from the CED-3 subfamily are known as initiator caspases in the apoptotic process.

All caspases share two important characteristics. First, the caspases are cysteine proteases containing a conserved QACXG pentapeptide, which contains the active-site cysteine. Second, these enzymes have a unique and strong preference for cleavage of the peptide bond C-terminal to aspartic acid

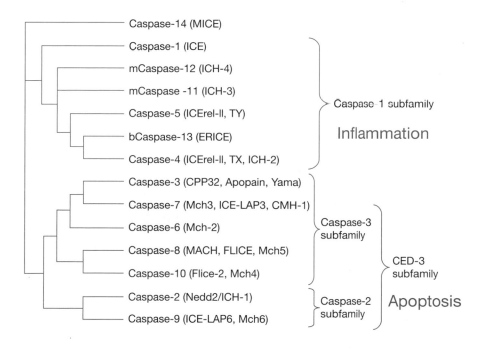

Figure 1. Caspases – cysteinyl aspartate-specific proteases
The phylogenic relationship of mammalian caspases appears to correlate with their function in inflammation or cell death. MICE, mini ICE; ICH, ICE/CED homologue; TY, transcript Y; TX, transcript X; ICErel, ICE-related protein; CPP32, cysteine protease protein of molecular mass 32 kDa; ICE-LAP, ICE-like apoptotic protease; CMH, CPP32/Mch 2 homologue; MACH, MORT-1-associated CED3 homologue; Flice, FADD-like ICE; mCaspase, mouse caspase; bCaspase, bovine caspase; Mch, mammalian CED homologue; NEDD, neuronally expressed developmentally down-regulated protein; ERICE, evolutionary related ICE.

residues. This primary cleavage specificity makes this family very rare among protease families. Of all known mammalian proteases, only the serine protease granzyme B has similar substrate specificity. Despite the requirement for aspartate at the substrate P1 site, the caspases differ in substrate specificity and can be divided into three subgroups based on preferred tetrapeptide sequences (P4–P1). The P4 site (four amino acids N-terminal to the cleavage site) is the most critical determinant of substrate specificity [8]. Caspases in group I (caspases 1, 4, 5 and 13) cleave preferentially after the (W/L)EHD motif, group II (caspases 3 and 7) after DEXD, whereas the optimal cleavage sequences for caspases belonging to group III (caspases 6, 8, 9 and 10) are (I/L/V)EXD. Only caspase-2 preferentially cleaves after the pentapeptide sequence VDVAD (Figure 2). The cleavage sites for caspases 11, 12 and 14 are not yet clear.

Activation of pro-caspases

Evidence for the sequential activation of caspases has led to the concept of a caspase cascade. This cascade begins with autocatalytic activation of initiator caspases that in turn transmit the signal by cleaving and thereby activating the downstream effector caspases [9]. As mentioned above, the initiator caspases (pro-caspases 2, 8, 9, 10 and 12), as well as the so-called inflammatory

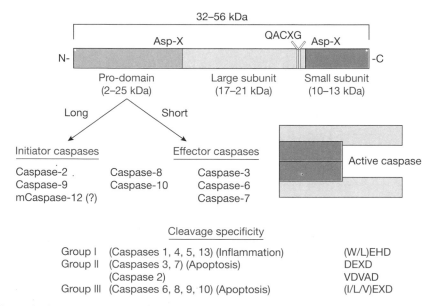

Figure 2. The structural and functional organization of caspases
The pro-caspases contain four domains: an N-terminal pro-domain, a large subunit, a small subunit and a short linker region between the large and small subunits. The active caspase is a tetramer composed of two large and two small subunits. Based on the size of pro-domains, pro-caspases can be subdivided into two groups: initiator and effector caspases. The caspases recognize a core tetrapeptide motif. Using a positional-scanning combinatorial substrate library it was possible to divide caspases into three groups with distinct cleavage specificity.

caspases (pro-caspases 1, 4, 5, 11 and 13), contain long pro-domains, while the pro-domains in effector caspases (pro-caspases 3, 6 and 7) are short. The long pro-domains contain distinct motifs, including DEDs (death effector domains) present in pro-caspases 8 and 10 and CARDs (caspase-recruitment domains) found in pro-caspases 1, 2, 4, 5 and 9, which are important for the activation of these enzymes [10]. Activation of caspases containing long pro-domains first requires oligomerization via DED or CARD domains. Following the recruitment of a single type of caspase pro-enzyme to a common oligomerization site, the low level of endogenous catalytic activity that is present in pro-enzymes is sufficient to initiate full catalytic activation by proteolysis next to the aspartic acid residue that is present at the junction between the large and small subunits (Figures 2 and 3). It is likely that pro-domain structure, rather than substrate specificity, determines the ability of pro-caspases to be activated by oligomerization. Active initiator caspases transmit the proteolytic signal by directly processing executioner, short-pro-domain-containing pro-caspases. This first cleavage by pre-existing active caspases is followed by autoproteolysis mediated by the low level of endogenous catalytic activity that is present in these effector pro-enzymes. It has been suggested that caspases can, in principle, undergo autocatalytic activation. However, the clear evidence for a non-recruitment type of autoactivation has not been presented.

There are two main intracellular pathways by which caspases can be activated; namely dependent on the ligation of death receptors or the release of apoptogenic factors from mitochondria (for a detailed review on apoptotic signalling, see [10,11]). Thus ligation of the TNF (tumour necrosis factor) or Fas/Apo 1/CD95 receptors results in the assembly of the so-called DISC (death-inducing signalling complex). Pro-caspase-8 via interaction with its DEDs is recruited to the DISC and becomes activated. In so-called type I cells, active caspase-8 cleaves and thereby activates pro-caspase-3, giving rise to the proteolytic cascade (Figure 4). In type II cells, caspase-8 cleaves the cytosolic protein Bid and its truncated form tBid translocates to mitochondria and induces the release of cytochrome *c*. It is important to note that caspase-8-

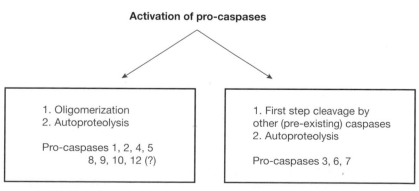

Figure 3. Activation of pro-caspases

mediated cleavage of Bid is not a unique mechanism of tBid formation. This protein can be cleaved by granzyme B and by lysosomal enzymes, as well as by calpain. In all cases translocation of tBid to the mitochondria was observed and was followed by the release of cytochrome *c* from the intermembrane space of mitochondria. Interestingly, in some experimental models calpain can also cleave caspases, and most often this cleavage inactivates caspase function. For example, calpain can cleave caspase-7 at sites distinct from those cleaved by the upstream caspases, generating proteolytically inactive fragments. Pro-caspases 8 and 9 can also be cleaved by calpains, and truncated caspase-9 is unable to activate pro-caspase-3. Finally, it has been reported that pro-caspase-3 cleavage by calpain results in the generation of a 29-kDa fragment, although it is unclear whether such cleavage causes the activation or inactivation of caspase-3 function (Figure 4). The best-known example of caspase activation by calpain is the cleavage of mouse pro-caspase-12, although the precise mechanism of this activation is unclear [12].

In most other apoptotic models, the caspases are activated downstream of mitochondrial release of cytochrome *c*. Upon entering the cytosol, cytochrome *c* binds to Apaf-1 (apoptotic protease-activating factor 1), which oligomerizes in the presence of dATP/ATP and binds pro-caspase-9 to form the so-called apoptosome complex [10]. This leads subsequently to autoproteolysis and activation of pro-caspase-9. In this scenario, again the activated initiator caspase

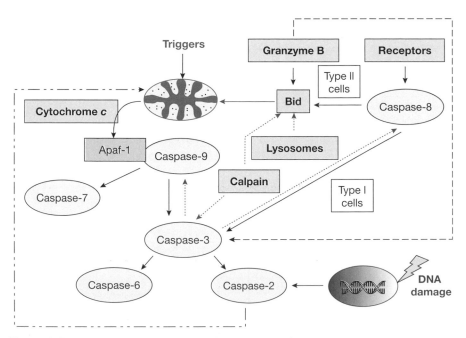

Figure 4. Signalling and caspase activation in apoptosis
Depending on the stimulus and cell type, several pathways of sequential caspase activation have been described. Different line shapes are related to distinct pathways of caspase activation.

(caspase-9), similar to caspase-8, propagates the death signal by proteolytically processing and activating downstream-located effector caspases (Figure 4).

Caspase-2 was one of the first mammalian apoptotic caspases to be identified [13] and its activity was implicated in cell death. However, the precise function of this enzyme in apoptosis was unclear. Recently, it has been shown that caspase-2 is required for DNA-damage-induced apoptosis [14]. It is the only pro-caspase that is present constitutively in the nucleus as well as in the Golgi apparatus. In response to DNA damage, activation of this enzyme was observed upstream of pro-caspase-9, indicating that caspase-2 acts as an effector of the mitochondrial pathway. Although the mechanism of pro-caspase-2 activation is unclear, the proposed model suggested that active caspase-2, or more likely a cleaved substrate, translocates out of the nucleus to target mitochondria and stimulates cytochrome c release. In this case, mitochondria may act as an amplifier rather than an initiator of caspase activity (Figure 4).

Intracellular inhibitors of caspase activity

A processed caspase is not necessarily catalytically active since processing and activation are under control of different protein factors, such as FADD (Fas-associated death domain) and Apaf-1, inhibitor proteins [FLIP (Flice-like inhibitory protein), IAPs (inhibitors of apoptosis protein) and Bcl-2-like proteins], HSPs (heat-shock proteins) and viral proteins [crmA, p35, vMIA (viral mitochondria-localized inhibitor of apoptosis)]. IAPs are a family of proteins that bind and inhibit caspases directly [15]. Studies have indicated that, unlike other caspases, pro-caspase-9 processing was necessary, but not sufficient, for catalytic activity [16]. Caspase-9 in complex with Apaf-1 seems to represent the active form of caspase-9. Thus Apaf-1 may not simply be an activator of caspase-9, but rather an essential regulatory subunit of the caspase-9 holoenzyme. Recruitment of pro-caspase-9 into the apoptosome complex and its activation are in addition controlled by HSPs and IAPs [17,18]. Activity of caspases is also controlled by several proteins, such as IAPs, crmA and HSPs. Recently, it has been shown that NO inhibits activity of at least seven caspases via nitrosation of the catalytic cysteine residue. In these experiments, *in vitro* caspase activity can be restored in the presence of dithiothreitol [19]. Interestingly, NO donors might also act directly at the level of the apoptosome and inhibit the sequential activation of caspases 9, 3 and 8 [20].

It has been suggested that Akt, a kinase that suppresses apoptosis, phosphorylates caspase-9, thereby preventing activation of this protease [21]. Despite the close similarity between human and mouse caspase-9, the latter does not contain an Akt phosphorylation site. The authors suggested that in mouse cells an additional evolutionarily conserved mechanism exists by which Akt can suppress activation of caspases in the mitochondrion-mediated pathway [21]. However, to prove this hypothesis, several critical experiments should be performed.

The caspases themselves are cysteine-dependent enzymes and, as such, appear to be redox-sensitive. One of the most reproducible inducers of caspase-mediated apoptosis is mild oxidative stress, although it is unclear how an oxidative stimulus can directly activate the caspase cascade. On the other hand, experiments with H_2O_2 suggested that prolonged or excessive oxidative stress can actually prevent caspase activation. A physiological example of this is the NADPH oxidase-derived oxidants generated by stimulated neutrophils that prevent caspase activation in these cells [22].

There are some indications for the possible role of ubiquitination in regulation of caspase activity; however, additional experiments are required to understand the significance of this observation.

Consequences of caspase activation during apoptosis

The pro-caspases are found in multiple intracellular compartments, including mitochondria, endoplasmic reticulum, Golgi apparatus, cytosol and nucleus, and the subcellular localization of the pro-enzyme is often different from the scene of action of the activated caspase. Several caspases, after being activated, translocate to other intracellular compartments, such as the endoplasmic reticulum and nucleus, where they cleave specific target proteins [23]. It is clear that if the proteolysis of substrate contributes functionally to the demise of the cell, this substrate should be cleaved prior to the biochemical and morphological manifestations of apoptotic cell death. In line with this hypothesis, caspase-mediated cleavage deregulates the activity of its cellular target prior to morphological changes (Figure 5).

Presently, more than 200 caspase substrates are known and the list continues to grow [24]. This list includes proteins, which carry out different

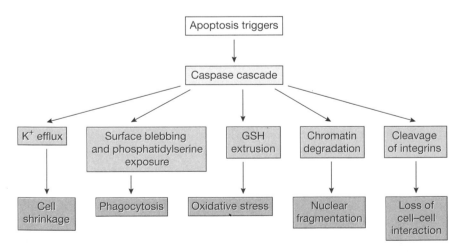

Figure 5. Some consequences of caspase activation during apoptosis
Apoptosis triggers initiate activation of the caspase cascade, which leads to the cleavage of target proteins and results in different biochemical and morphological manifestations of cell death.

functions. Among these proteins are structural proteins, signalling proteins, regulators of transcription and translation, regulators of replication and cell cycle, regulators of DNA repair and cleavage, regulators of RNA metabolism, regulators of cell–cell interaction, regulators of pro-inflammatory cytokines, regulators of apoptosis and many others. Thus caspase-mediated cleavage can result in either inactivation (in most cases) of proteins essential for the cell's structural integrity or survival, or activation of dormant, pro-apoptotic proteins. Among the proteins that are inactivated by cleavage are cytoskeletal proteins, including lamin, α-fodrin and actin, and proteins involved in DNA repair and cell-cycle regulation, such as PARP (polyADP-ribose polymerase) and retinoblastoma protein, respectively. In addition to the pro-caspases themselves, caspase substrates that are activated upon cleavage of an inhibitory domain include protein kinase Cδ and transcription factor sterol regulatory element-binding proteins. The CAD (caspase-activated DNase) is indirectly activated by cleavage of an inhibitory subunit, ICAD (inhibitor of CAD). This caspase-mediated cleavage results in release and activation of the catalytic subunit of CAD (Figure 6).

Although caspases do not contain a nuclear localization signal, many of these enzymes are translocated into the nuclei of apoptotic cells [25]. Numerous nuclear-located proteins are cleaved by different caspases, suggesting that caspase translocation into the nucleus is an important event during apoptosis (Figure 6).

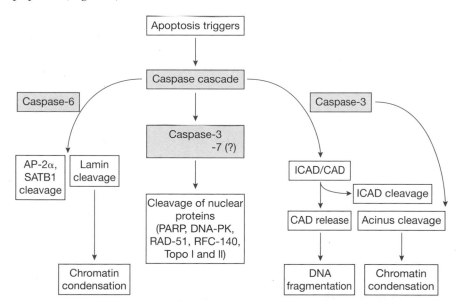

Figure 6. Role of caspases in nuclear apoptosis
Translocation of active caspases into the nucleus results in cleavage of many nuclear proteins, which leads to DNA fragmentation, chromatin margination and nuclear collapse. CAD, caspase-activated DNase; ICAD, inhibitor of CAD; AP-2α, activator protein-2α; SATB1, special AT-rich sequence-binding protein 1. RAD, protein involved in homologous recombination and DNA repair; RFC, replication factor C; DNA-PK, DNA-dependent protein kinase.

Caspase-3 is activated during most apoptotic processes and is believed to be the main executioner caspase. The presence of active caspase-3 in the nucleus is essential for cleavage of different proteins, for DNA fragmentation and chromatin condensation. Caspase-7 cleaves PARP and some other nuclear proteins *in vitro*; however, clear evidence for its activity in the nuclei of apoptotic cells is still absent. Caspase-6 has been shown by many groups to cleave lamin. However, in *casp-6*-null cells and in *casp-6*-knockout mice, the cleavage of lamin was observed in response to diverse stimuli [26]. In addition to lamin, caspase-6 cleaves several nuclear proteins, such as transcription factor activator protein-2α, SATB1 (special AT-rich sequence-binding protein 1) and the viral nucleocapsid protein of transmissible gastroenteritis coronavirus. Interestingly, experiments using cell-free extracts immuno-depleted of caspase-3, -6 or -7 have shown that depletion of caspase-3 results in the absence of cleavage of a majority of nuclear substrates as well as DNA fragmentation, chromatin marginization and nuclear collapse [27]. However, although caspases 6 and 7 are considered to be important downstream effector caspases, depletion of either caspase had minimal impact on any of the nuclear parameters investigated, raising questions concerning their precise role during the execution step of apoptosis within the nucleus, at least in cell-free extracts. Although active caspases 2 and 9 were detected in the nuclei of apoptotic cells, their precise targets within this intracellular compartment are unknown.

Even though caspase-independent cell death exists, most experimental models of apoptosis involve the activation of caspases. Until now, no activation of caspases has been found during necrotic cell death. Moreover, inactivation of caspases might switch the death process from apoptosis to necrosis. Therefore, the detection of activated caspases can be used as a biochemical marker of apoptosis induced by diverse stimuli in many types of cells.

Methods for analysis of caspase identification or activation

Detection and identification of different caspases was performed using two-dimensional PAGE [28]. In two-dimensional PAGE, proteins are separated according to their isoelectric point and size. Since the various caspases have different isoelectric points and sizes, they will appear in different areas on the two-dimensional gel.

Determination of caspase activation can be performed in different ways and the advantages and disadvantages of each technique were outlined recently [29].

Detection of activated caspases by immunoblotting

As mentioned above, pro-caspases are proteolytically cleaved to become active enzymes. This processing of a particular pro-caspase can be studied by using the immunoblot technique. The appearance of the small or large subunit of a mature caspase can be detected with a specific antibody. The cleavage of a pro-caspase with the appearance of large and small subunits can be used as a marker for processing/activation of the specific caspase. However, as discussed above, the processed caspase is not necessarily activated. Therefore, to confirm that processed caspases are also catalytically active, additional studies on enzymic activity using synthetic peptides or known caspase target proteins might be performed.

Cleavage of synthetic substrates

One of the commonly used ways to detect caspase activity is to measure the cleavage of synthetic substrates upon their incubation with lysates of apoptotic cells. The principle of this technique was described several years ago [30]. Substrates containing tetrapeptide sequences mimicking the cleavage sites of the various caspase subgroups have been synthesized and are now commercially available. The tetrapeptides are conjugated to either the chromophore *p*-nitroanilide or a fluorochrome, such as 7-amino-4-methylcoumarin or 7-amino-4-trifluoromethyl coumarin. Upon cleavage of the substrate, the liberated chromophore *p*-nitroanilide or fluorophores 7-amino-4-methylcoumarin and 7-amino-4-trifluoromethyl coumarin are analysed by spectrometry or fluorometry, respectively. This method can be performed using whole-cell extracts, complex mixtures or purified caspases.

To determine the activity of a specific caspase and not a subgroup of caspases, a fluorometric immunosorbent enzyme assay has been developed. In this assay the caspase of interest is captured from cell lysates using a monoclonal antibody. After washing, the tetrapeptide substrate is added and the caspase activity is measured as described above.

Caspase inhibitors and affinity labelling

Caspase inhibitors can be used to determine the importance of caspases for apoptotic cell death in general or for a certain apoptotic event in particular. In systems where the activation of caspases is crucial for apoptosis to occur, the inhibitors block or at least delay cell death. Caspase inhibitors act by binding to the active site of caspases and form either a reversible or an irreversible linkage. Many of the irreversible inhibitors function by forming covalent adducts with the cysteine in the active site of the caspase [7]. Peptide inhibitors that have preferences for specific caspases, or caspase subgroups, include a caspase-recognition sequence (often the same as the preferred tetrapeptide sequence for substrate cleavage) conjugated to a functional group such as an aldehyde, chloromethylketone, FMK (fluoromethylketone) or fluoroacyloxymethylketone. Caspase inhibitors with an aldehyde group are reversible and those with chloromethylketone, FMK or fluoroacyloxymethylketone groups are irreversible inhibitors. To determine the activation of a particular caspase, a specific inhibitor, such as biotinylated *N*-acetyl-Asp-DEVD-aldehyde, can be employed that preferentially labels active caspase-3-like proteases.

The ELISA is another method that can be recommended for determination of the activation of a specific caspase when using biotin-labelled inhibitors. After incubation with a biotinylated inhibitor, which labels active caspases in a sample, a specific antibody is added in order to capture the caspase of interest. The antibody-bound caspase is then detected with horseradish peroxidase-conjugated streptavidin that binds to biotin. Although the antibody captures both active and latent enzyme, the biotinylated inhibitor binds only to the active enzyme and the two are thereby distinguished.

Detection of cleavage of caspase substrates by immunoblotting

Many proteins are specifically cleaved by caspases during the apoptotic process. Hence, the identification of these protein fragments allows one to judge indirectly the activation of caspases. The proteins in a sample are resolved by SDS/PAGE and transferred to a membrane by the immunoblot procedure. Many antibodies that can be used for detection of cellular caspase substrates by this technique are commercially available.

Activation of caspases in intact cells investigated by confocal or fluorescence microscopy and/or flow cytometry

The activation of caspases can also be detected in whole cells using several approaches. PhiPhiLux is a cell-permeable fluorogenic caspase substrate that can be used to detect and measure the activation of caspases in cells. PhiPhiLux contains the amino acid sequence DEVDGI that is recognized and cleaved by caspase-3-like enzymes (caspases in subgroup II). In addition to the caspase-recognition motif, the substrate contains two rhodamines (which fluoresce green) or rhodamine derivatives (which fluoresce red). The fluorescence is quenched in the non-cleaved substrate, but not in the cleaved ones. Hence, the fluorescence observed in cells incubated with the substrate corresponds to the activity of caspases belonging to subgroup II and can be detected by flow cytometry or fluorescence microscopy.

Activation of caspases in intact cells can also be assayed by immunohistochemistry using antibodies against either the active caspase or any caspase-cleaved products. The primary antibody as well as a secondary antibody can be conjugated to a fluorescent probe, such as FITC (fluorescein isothiocyanate). In this situation, the stained cells can be analysed by flow cytometry and by fluorescent or laser scanning confocal microscopy.

Proteins cleaved by caspases can also be detected using immunohistochemistry. For instance, antibodies against cleaved PARP, actin and cytokeratin 18 have been used extensively for this purpose. In addition to tissue sections, immunostaining can also be performed in the analysis of cultured cells (immunocytochemistry). Adherent cells can be cultured directly on coverslips and cells growing in suspension can be sedimented on to a coverslip using the cytospin technique.

Recently, a new approach using recombinant caspase substrates for monitoring caspase activity *in vivo* has been developed [31]. The substrate contained tandem repeats of GFP (green fluorescent protein) separated by DEVD or other cleavage sites. Cells were transfected with a vector containing the gene encoding the substrate using a tetracycline-inducible expression system. The expressed caspase substrate was cleaved upon apoptosis induction. Cleavage of the substrate releases GFP and this prevents the fluorescence resonance energy transfer, which is a phenomenon that occurs when two fluorophores with overlapping emission and excitation spectra, such as GFP, are physically close to each other [32]. The increase in fluorescence upon cleavage of the caspase substrate can be detected using a fluorimeter or flow cytometer as well as with fluorescent or laser-scanning confocal microscopes.

CaspaTags and FLICAs (fluorochrome-labelled inhibitors of caspases) are newly developed types of fluorescent probe suggested for use in the detection of activated caspases in living cells. These compounds are carboxyfluorescein-labelled FMK inhibitors, which are cell-permeant and should bind irreversibly to the active site of individual caspases (containing the caspase-recognition sequence for each caspase separately) or to all caspases (containing the amino acid sequence VAD) [33]. It has been shown that CaspaTags and FLICAs exclusively stain cells containing the processed caspases of interest and the probes do not accumulate in normal (non-apoptotic) cells. Therefore, apoptotic cells containing active caspases should be identified and distinguished from non-apoptotic cells by their fluorescence using flow cytometry, fluorescence microscopy or a fluorescence microtitre plate reader. However, prior to the publication of this volume, additional control experiments were performed by the authors and results have shown that a predominant component in the binding of these reagents to apoptotic cells does not reflect their direct reactivity with the protease's active centres. Moreover, the bulk portion of fluorochrome-tagged protease inhibitors that react with apoptotic cells bind to sites other than the active enzyme centres of the respective enzymes (Z. Darzynkiewicz, G. Jahnson and B.W. Lee, personal communication submitted for distribution via the International Cell Death Society). Thus the labelling of apoptotic cells with these reagents is not a specific marker of caspase activation and therefore should be taken into consideration.

Given the critical importance of caspases in the realization of the apoptotic process, development of new more sensitive methods of measurement of caspase activity *in vivo* has remained high on the wish list of apoptosis researchers.

Conclusions

Activation of caspase family proteases has been detected in numerous cell systems and appears to function as a pathway through which apoptotic mechanisms operate. Our understanding of caspase activation and effector function has been greatly enhanced during the last 10 years; however, a lot of ground still remains to be covered. For example, the precise role of several caspases in cell death, as well as mechanisms of their activation or regulation of their function is still unclear. The functional significance of the majority of cleaved caspase substrates also remains to be determined. Since inappropriate apoptosis is clearly associated with the aetiology of several human diseases [34] and the role of caspase in the regulation of cell death has been well documented, the manipulation of caspase activation/activity might be an important key in the treatment of these diseases.

Summary

- *The caspase family of proteases consists of at least 14 mammalian members. A phylogenic analysis revealed that this gene family is composed of two major subfamilies, which are either involved in the processing of cytokines or play a central role in the initiation and execution phases of cell death.*
- *All caspases share two important characteristics. First, the caspases are cysteine proteases containing a conserved QACXG pentapeptide, which contains the active-site cysteine. Second, these enzymes have a unique strong preference for cleavage of the peptide bond C-terminal to aspartate.*
- *Processing, activation and activity of all caspases are tightly regulated by different protein factors.*
- *The sequential activation of caspases has led to the concept of a caspase cascade. Caspase-mediated cleavage deregulates the activity of its cellular target prior to morphological changes.*
- *Detection of activated caspases can be used as a biochemical marker of apoptosis induced by diverse stimuli in many types of cells.*

I thank the Swedish Cancer Foundation, the Stockholm Cancer Society, the Swedish Medical Research Council and an EC-RTD program for support of work which led to some of the findings and concepts described in this chapter.

References

1   Yuan, J.S., Shaham, S., Ledoux, S., Ellis, H.M. & Horvitz, H.R. (1993) The *C. elegans* cell death gene ced-3 encodes a protein similar to mammalian interleukin-1β-converting enzyme. *Cell* **75**, 641–652

2   Alnemri, E.S., Livingston, D.J., Nicholson, D.W., Salvesen, G., Thornberry, N.A., Wong, W.W. & Yuan, J. (1996) Human ICE/CED-3 protease nomenclature. *Cell* **87**, 171

3   Fadeel, B., Orrenius, S. & Zhivotovsky, B. (2000) The most unkindest cut of all: on the multiple roles of mammalian caspases. *Leukemia* **14**, 1514–1525

4   Nicholson, D.W. (1999) Caspase structure, proteolytic substrates, and function during apoptotic cell death. *Cell Death Differ.* **6**, 1028–1042

5   Lamkanfi, M., Declercq, W., Kalai, M., Saelens, X. & Vandenabeele, P. (2002) Alice in caspase land. A phylogenetic analysis of caspases from worm to man. *Cell Death Differ.* **9**, 358–361

6   Earnshaw, W.C., Martins, L.M. & Kaufmann, S.H. (1999) Mammalian caspases: structure, activation, substrates, and functions during apoptosis. *Annu. Rev. Biochem.* **68**, 383–424

7   Lippens, S., Kockx, M., Knaapen, M., Mortier, L., Polakowska, R., Verheyen, A., Garmyn, M., Zwijsen, A., Formstecher, P., Huylebroeck, D. et al. (2000) Epidermal differentiation does not involve the pro-apoptotic executioner caspases, but is associated with caspase-14 induction and processing. *Cell Death Differ.* **7**, 1218–1224

8   Thornberry, N.A., Rano, T.A., Peterson, E.P., Rasper, D.M., Timkey, T., Garcia-Calvo, M., Houtzager, V.M., Nordstrom, P.A., Roy, S., Vaillancourt, J.P. et al. (1997) A combinatorial approach defines specificities of members of the caspase family and granzyme B. *J. Biol. Chem.* **272**, 17907–17911

9   Stennicke, H.R. & Salvesen, G.S. (1999) Catalytic properties of the caspases. *Cell Death Differ.* **6**, 1054–1059

10   Budihardjo, I., Oliver, H., Lutter, M., Luo, X. & Wang, XD. (1999) Biochemical pathways of caspase activation during apoptosis. *Annu. Rev. Cell Devel. Biol.* **15**, 269–290

11   Hengartner, M.O. (2000) The biochemistry of apoptosis. *Nature (London)* **407**, 770–776

12   Nakagawa, T. & Yuan, J. (2000) Cross-talk between two cysteine protease families. Activation of caspase-12 by calpain in apoptosis. *J. Cell Biol.* **150**, 887–894

13   Kumar, S., Kinoshita, M., Noda, M., Copeland, N.G. & Jenkins, N.A. (1994) Induction of apoptosis by the mouse Nedd2 gene, which encodes a protein similar to the product of the *Caenorhabditis elegans* cell death gene ced-3 and the mammalian IL-1β-converting enzyme. *Genes Dev.* **8**, 1613–1626

14   Robertson, J.D., Enoksson, M., Suomela, M., Zhivotovsky, B. & Orrenius, S. (2002) Caspase-2 acts upstream of mitochondria to promote cytochrome c release during etoposide-induced apoptosis. *J. Biol. Chem.* **277**, 29803–29809

15   Deveraux, Q.L. & Reed, J.C. (1999) IAP family proteins-suppressors of apoptosis. *Genes Dev.* **13**, 239–252

16   Rodriguez, J. & Lazebnik, Y. (1999) Caspase-9 and APAF-1 form an active holoenzyme. *Genes Dev.* **13**, 3179–3184

17   Beere, H.M., Wolf, B.B., Cain, K., Mosser, D.D., Mahboubi, A., Kuwana, T., Tailor, P., Morimoto, R.I., Cohen, G.M. & Green, D.R. (2000) Heat-shock protein 70 inhibits apoptosis by preventing recruitment of procaspase-9 to the Apaf-1 apoptosome. *Nat. Cell Biol.* **2**, 469–475

18   Bratton, S.B., Walker, G., Srinivasula, S.M., Sun, X.M., Butterworth, M., Alnemri, E.S. & Cohen, G.M. (2001) Recruitment, activation and retention of caspases-9 and -3 by Apaf-1 apoptosome and associated XIAP complexes. *EMBO J.* **20**, 998–1009

19   Melino, G., Bernassola, F., Knight, R.A., Corasaniti, M.T., Nistico, G. & Finazzi-Agro, A. (1997) S-nitrosylation regulates apoptosis. *Nature (London)* **388**, 432–433

20   Zech, B., Köhl, R., von Knethen, A. & Brüne, B. (2003) Nitric oxide donors inhibit formation of the Apaf-1/caspase-9 apoptosome and activation of caspases. *Biochem. J.* **371**, 1055–1064

21   Cardone, M.H., Roy, N., Stennicke, H.R., Salvesen, G.S., Franke, T.F., Stanbridge, E., Frisch, S. & Reed, J.C. (1998) Regulation of cell death protease caspase-9 by phosphorylation. *Science* **282**, 1318–1321

22   Hampton, M.B. & Orrenius, S. (1999) Dual regulation of caspase activity by hydrogen peroxide: implications for apoptosis. *FEBS Lett.* **414**, 552–556

23   Zhivotovsky, B., Samali, A., Gahm, A. & Orrenius, S. (1999) Caspases: their intracellular localization and translocation during apoptosis. *Cell Death Differ.* **6**, 644–651

24   Fischer, U., Jännicke, R.U. & Schulze-Osthoff, K. (2003) Many cuts to ruin: a comprehensive update of caspase substrates. *Cell Death Differ.* **10**, 76–100

25   Robertson, J.D., Orrenius, S. & Zhivotovsky, B. (2000) Nuclear events in apoptosis. *J. Struct. Biol.* **129**, 346–358

© 2003 The Biochemical Society

26   Zheng, T.S. & Flavell, R.A. (2000) Divinations and surprises: genetic analysis of caspase function in mice. *Exp. Cell Res.* **256**, 67–73

27   Slee, E.A., Adrain, C. & Martin, S.J. (2001) Executioner caspase-3, -6, and -7 perform distinct, non-redundant roles during the demolition phase of apoptosis. *J. Biol. Chem.* **276**, 7320–7326

28   Faleiro, L., Kobayashi, R., Fearnhead, H. & Lazebnik, Y. (1997) Multiple species of CPP32 and Mch2 are the major active caspases present in apoptotic cells. *EMBO J.* **16**, 2271–2281

29   Kohler, C., Orrenius, S. & Zhivotovsky, B. (2002) Evaluation of caspase activity in apoptotic cells. *J. Immunol. Methods* **265**, 97–110

30   Pennington, M.W. & Thornberry, N.A. (1994) Synthesis of a fluorogenic interleukin-1β converting enzyme substrate based on resonance energy transfer. *Pept. Res.* **7**, 72–76

31   Tawa, P., Tam, J., Cassady, R., Nicholson, D.W. & Xanthoudakis, S. (2001) Quantitative analysis of fluorescent caspase substrate cleavage in intact cells and identification of novel inhibitors of apoptosis. *Cell Death Differ.* **8**, 30–37

32   Pollok, B.A. & Heim, R. (1999) Using GFP in FRET-based applications. *Trends Cell Biol.* **9**, 57–60

33   Bedner, E., Smolewski, P., Amstad, P. & Darzynkiewicz, Z. (2000) Activation of caspases measured *in situ* by binding of fluorochrome-labeled inhibitors of caspases (FLICA): correlation with DNA fragmentation. *Exp. Cell. Res.* **259**, 308–313

34   Fadeel, B., Orrenius, S. & Zhivotovsky, B. (1999) Apoptosis in human disease: A new skin for the old ceremony? *Biochem. Biophys. Res. Commun.* **266**, 699–717

4

Apoptosis: bombarding the mitochondria

Philippe Parone, Muriel Priault, Dominic James, Steven F. Nothwehr and Jean-Claude Martinou[1]

Department of Cell Biology, University of Geneva, 30 quai Ansermet, 1211 Geneva 4, Switzerland

Abstract

Mitochondria play a central role in apoptosis triggered by many stimuli. They integrate death signals through Bcl-2 family members and co-ordinate caspase activation through the release of apoptogenic factors that are normally sequestered in the mitochondrial intermembrane space. The release of these proteins is the result of the outer mitochondrial membrane becoming permeable. In addition, mitochondria can initiate apoptosis through the production of reactive oxygen species.

Introduction

Mitochondria are organelles that serve a variety of cellular functions, including ATP production, by aerobic metabolism, ionic homoeostasis maintenance, oxidation of carbohydrates and fatty acids, and the regulation of apoptosis. Their dysfunction has been proposed to be involved in several diseases including degenerative disorders. In this chapter, we will emphasize the role played by mitochondria in apoptosis, through the release of apoptogenic factors and the production of free radicals.

[1]*To whom correspondence should be addressed (e-mail Jean-claude.martinou@cellbio.unige.ch).*

Mitochondria and apoptosis: some history

Mitochondria have been known for a long time to cause cellular damage and cell death through the production of free radicals [1]. However, it is only since the anti-apoptotic protein Bcl-2 (which stands for B-cell lymphoma 2) was found to reside in this organelle [2] that scientists started to consider mitochondria as important players in apoptosis and envisioned a mitochondrial function for Bcl-2. Two main findings reinforced the existence of a connection between Bcl-2, mitochondria and apoptosis: first, Bcl-2 was found to act as an antioxidant protein [3] and secondly it was reported that TNF (tumour necrosis factor)-induced apoptosis of L929 cells was accompanied by a drop in mitochondrial membrane potential ($\Delta\Psi$) that was prevented by Bcl-2 over-expression [4]. The occurrence of a drop in $\Delta\Psi$ in many cell types undergoing apoptosis was then confirmed by a plethora of laboratories [5]. In the meantime, Newmeyer et al. [6] discovered that a mitochondrial factor was required for the activation of caspases (cysteinyl aspartate-specific proteases) [6]. This factor was later identified as cytochrome c, an electron transporter of the mitochondrial respiratory chain which was found to be released from mitochondria during apoptosis [7]. This result was consistent with previous data showing that mitochondrial proteins are released during apoptosis [8], an observation previously reported to be the consequence of the opening of a mitochondrial megachannel called the permeability transition pore [9]. The link between opening of this channel and the release of mitochondrial proteins during apoptosis was rapidly made, rightly or wrongly, as discussed below.

Apoptotic pathways engaging mitochondria

Two main apoptotic pathways have been described over the past few years: the death-receptor pathway and the mitochondrial pathway [10]. The first is engaged by death receptors such as TNF or Fas receptors, which, upon binding to their appropriate ligands, form a death-inducing signalling complex. This complex is formed by association of their cytoplasmic death domain, the adaptor protein FADD (Fas-associated death domain) and pro-caspase-8, which is proteolytically cleaved to generate the active enzyme. In type I cells such as T-lymphocytes, caspase-8 then cleaves downstream caspases such as caspases 3 or 7 that execute the cell. In type II cells, such as hepatocytes, caspase-8 cleaves Bid, a BH3 (Bcl-2 homology 3)-only protein whose C-terminal fragment translocates to mitochondria to engage the mitochondrial pathway. Independently of death-receptor activation, the mitochondrial pathway can also be activated in response to a large number of death stimuli including DNA damage, topoisomerase inhibition or trophic-factor deprivation. This process culminates in the release of mitochondrial proteins from the intermembrane space into the cytosol (Figure 1).

What are these mitochondrial proteins whose delocalization in the cytosol contributes to apoptosis and what are their mechanisms of action?

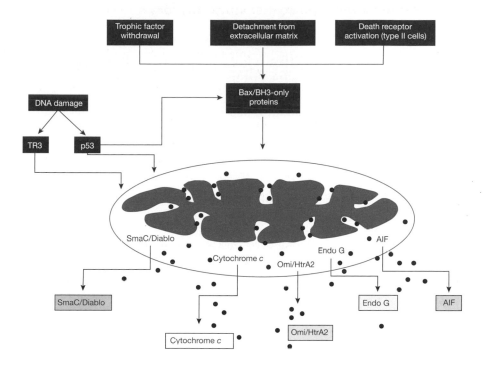

Figure 1. Release of mitochondrial apoptogenic factors

Many death stimuli can trigger the mitochondrial pathway that involves Bcl-2 family members. Upon activation by BH3-only proteins, pro-apoptotic members of the Bcl-2 family such as Bax render the outer mitochondrial membrane permeable. Other proteins such as p53 can directly neutralize anti-apoptotic members of the Bcl-2 family. As a consequence, many proteins that are normally sequestered in the intermembrane space of mitochondria, including cytochrome c, Smac/Diablo, Omi/HtrA2, AIF and Endo G (endonuclease G), are released into the cytosol.

The first such protein to be identified was cytochrome c. Using a HeLa cell-free system, Wang's group [7] showed that induction of caspase activity by addition of dATP was dependent on the presence of cytochrome c released during the preparation of the cytosolic extract. Upon dATP and cytochrome c binding, Apaf-1 (apoptotic protease-activating factor 1) undergoes conformational changes, forms a heptamer and recruits and activates pro-caspase-9 [11].

In addition to cytochrome c, a protein called Smac/Diablo [second mitochondrial activator of caspases/direct IAP (inhibitor of apoptosis protein)-binding protein] [12] was also found to be released. In the cytosol, Smac/Diablo binds to and inhibits IAPs (inhibitors of apoptosis) [13], which are the physiological caspases 3 and 9 inhibitors, and thereby contributes to the activation of caspases. Another mitochondrial protein displaying a similar action is Omi/HtrA2 [14], a serine protease that is another IAP inhibitor. AIF (apoptosis-inducing factor) [15], a flavoprotein with redox activity, and endonuclease G [16], a protein thought to be a mitochondrial matrix protein involved in mitochondrial DNA replication, are two other proteins that, once released from mitochondria, appear to translocate to the nucleus where they contribute to the cleavage of DNA. The release of AIF and endonuclease G, in contrast to the release of cytochrome *c*, Smac/Diablo and Omi/HtrA2, appears to be dependent on the activation of caspases, as shown recently in *Caenorhabditis elegans*, suggesting that they may be involved in the late events of apoptosis [17].

The release of all these apoptogenic factors can only occur because the outer mitochondrial membrane becomes leaky. The mechanisms that lead to permeabilization of the outer mitochondrial membrane are controlled by Bcl-2 family members.

Bcl-2 family members regulate the permeability of the outer mitochondrial membrane

Bcl-2 family members are composed of a group of anti-apoptotic proteins (e.g. Bcl-2 and Bcl-x_L) and a group of pro-apoptotic proteins that display three BH domains, BH 1, 2 and 3 (Bax, Bak and Bok) or only one BH3 domain (BH3-only proteins, e.g. Bid and Bim) [18]. The activity of the BH3-only proteins appears to be regulated at the transcriptional or post-translational levels (Figure 2).

Transcriptional regulation

Egl1 is a BH3-only protein expressed in *C. elegans*. Its expression is regulated by various transcription factors including Tra-1A, Ces-1 and Ces-2 [19]. In mammals, the BH3-only proteins Noxa and Puma have been shown to be transcriptionally up-regulated in response to p53, a transcription factor activated by DNA-damaging agents [20]. Other BH3-only proteins, Bim and Hrk, have been also shown to undergo regulation at the transcriptional level in response to activation of the Jun kinase pathway in trophic-factor-deprived neurons [21].

Sequestration by the cytoskeleton

The regulation of Bim appears complex since another mode of regulation has been described for this protein. In addition to transcriptional regulation, BH3-only proteins can be kept inactive by sequestration by the cytoskeleton. Bim is produced as three alternatively spliced products from the same gene (Bim EL, Bim L and Bim S). Bim S been the most potent death inducer whereas Bim EL and Bim L are less apoptotic because they can be trapped by microtubules through binding with the light chain of dynein LC8, a component of the dynein motor complex [22]. In cells undergoing apoptosis, Bim EL and Bim L detach from the microtubules by a mechanism that is not well understood and appear to interact with the dynein light chain 1 of the actin-cytoskeleton-based myosin V motor complex [22].

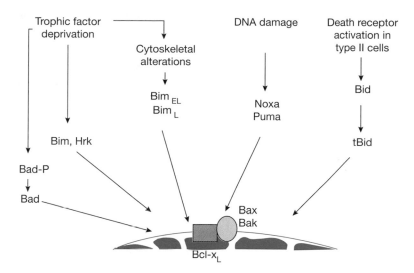

Figure 2. Mechanisms leading to activation of BH3-only proteins
BH3-only proteins can be activated transcriptionally (Noxa, Puma, Bim, Hrk), by cleavage (Bid), by dephosphorylation (Bad) or following remodelling of the cytoskeleton (Bim, Bmf). Once activated these proteins either inhibit anti-apoptotic members of the Bcl-2 family, such as Bcl-x$_L$, or activate pro-apoptotic members, such as Bax.

Limited proteolysis and phosphorylation

Bid is activated following cleavage by caspase-8 as described above. Bid cleavage appears to be regulated by protein kinase CKI- and CKII-mediated phosphorylation at the Ser-61 and Ser-64 residues of Bid. Phosphorylation renders the protein less sensitive to cleavage by caspase-8 [23]. This suggests that following activation of death receptors in type II cells, Bid must be dephosphorylated to be cleaved. Bad activity is down-regulated through phosphorylation at different sites by various kinases, including Akt. When phosphorylated, Bad is sequestered by the 14-3-3 scaffold protein and cannot interact with Bcl-2 or Bcl-x$_L$. Following dephosphorylation, as in the case of trophic-factor deprivation, Bad would interact with anti-apoptotic proteins thereby inhibiting their function. The number and the type of BH3-only proteins that intervene in response to a given cell death stimulus is unknown. It is possible that for each death stimulus, whether it is mediated by plasma-membrane receptors (death receptors, for example) or from inside the cell (for example, following DNA damage or dysfunction of an organelle), a battery of specific BH3-only proteins posted at key regions in the cell would selectively intervene to activate the cell-death programme. This would explain why there are so many proteins of this type. Once activated, BH3-only proteins would either interact with and inhibit anti-apoptotic proteins (the case for Bad or Bim) or interact with and activate the pro-apoptotic proteins such as Bax or Bak (the case for Bid). Apoptosis would occur only if the anti-apoptotic proteins are inhibited and, at the same time, the pro-apoptotic proteins activated [24]. The model would also predict that inhibition of anti-apoptotic proteins, without concomitant activation of the pro-apoptotic proteins, would sensitize cells to death stimuli.

Besides BH3-only proteins, recent studies have also shown that two transcriptional activators, p53 and TR3, an orphan receptor from the steroid/thyroid receptor family, have been shown to relocate from the nucleus to the surface of the mitochondria. There they trigger the release of cytochrome c, possibly by directly interacting with Bcl-2 or Bcl-x_L as it has been shown for p53 [25].

In summary, it follows from these studies that the mechanisms leading to activation of pro-apoptotic members such as Bax represent a key event during apoptosis.

Bax activation during apoptosis

In healthy cells, Bax is normally cytosolic, and translocates to mitochondria during apoptosis. Bax appears to be kept inactive in the cytosol thanks to a protein, Ku70, that binds to its N-terminus [26]. In addition to Ku70, Humanin, a protein that is up-regulated in the brain of patients with Alzheimer's disease, also binds to Bax and keeps it inactive in the cytosol [27]. In cells undergoing apoptosis, the dissociation of these inhibitory proteins from Bax would constitute the first of a series of steps leading to Bax activation. This translocation is probably the consequence of conformational changes that result in the exposure of the N- and C-termini of the protein. The three-dimensional structure of Bax revealed that the putative C-terminal transmembrane domain (helix $\alpha9$) is masked inside the hydrophobic core of the protein that consists of eight amphipathic helices clustered around helix $\alpha5$ [28]. The helix $\alpha9$ acts as a mitochondrial localization signal since its fusion to the C-terminus of green fluorescent protein is sufficient to target green fluorescent protein to mitochondria. In order to target Bax to the outer mitochondrial membrane, the helix $\alpha9$ would have to disengage from the hydrophobic pocket. What triggers the conformational change of Bax is still unknown. In addition to BH3-only proteins, such as tBid, other factors are probably required, which we think form a complex with cytosolic factors and at least one mitochondrial protein acting as a Bax-docking site on the outer mitochondrial membrane [29]. Once inserted in the outer mitochondrial membrane or prior to insertion, Bax oligomerizes and triggers the permeabilization of the outer mitochondrial membrane by an as-yet-unknown mechanism, as described later [10]. One of the mechanisms of action of the anti-apoptotic members of the Bcl-2 family, such as Bcl-2 or Bcl-x_L, is to prevent Bax insertion and oligomerization. Interestingly, these proteins display their anti-apoptotic action when targeted exclusively to the endoplasmic reticulum. In cells overexpressing Bcl-2, the levels of calcium in the endoplasmic reticulum were lower than normal [30]. How these proteins may regulate calcium homoeostasis of the endoplasmic reticulum and influence mitochondrial dysregulation and apoptosis remains to be clarified.

Theories of outer-mitochondrial-membrane permeabilization

The mechanisms whereby pro-apoptotic members of the Bcl-2 family trigger cytochrome c release remain elusive. Several models have been postulated [10].

Rupture of the outer mitochondrial membrane

According to this theory, water and solutes enter the matrix during apoptosis, causing swelling of the mitochondria. Because the inner membrane has a considerably larger surface area than the outer membrane, expansion of the inner membrane upon matrix swelling can break the outer membrane. This would cause passive efflux of the whole contents of the intermembrane space into the cytosol. Two models can account for matrix swelling. The first implicates the initial hyperpolarization of the inner membrane that would result from the inability to exchange mitochondrial ATP for cytosolic ADP during apoptosis. This antiport is normally mediated by the VDAC (voltage-dependent anion channel) located in the outer mitochondrial membrane and the ANT (adenylate translocator), which resides in the inner membrane. The hyperpolarization of the inner membrane is predicted to result in an osmotic matrix swelling. The second model postulates opening of the PTP (permeability transition pore). This is a high-conductance channel that is mainly formed by the apposition of the VDAC and ANT channels at contact sites between the outer and inner mitochondrial membranes. Opening of the PTP would trigger a sudden increase in the permeability of the outer membrane to molecules of <1.5 kDa. This event would result in a drop in membrane potential and osmotic swelling of the matrix. In this model, the channel would open following interaction of Bax with ANT.

Cytochrome c-conducting channels

A major breakthrough in the understanding of the mechanisms of action of Bcl-2 family members has been the discovery that the structure of Bcl-x_L, Bax and Bid is reminiscent of the structure of the transmembrane domain of the bacterial colicin toxins and diphtheria toxin. Because the transmembrane domain of these toxins has been shown to form a pore, it was hypothesized that Bcl-2-family proteins could also be pore-forming proteins. This was confirmed for Bax, Bcl-x_L and Bid, which form channels of various conductances across synthetic lipid membranes, the largest ones being formed by Bax. These data suggest that a Bax oligomer (maybe a tetramer), makes a pore large enough to allow the release of mitochondrial apoptogenic factors. Anti-apoptotic proteins could either prevent the formation of the Bax pore or modify the structure of the pore such that it would be unable to allow the efflux of mitochondrial proteins. Another possibility is that Bax may interact with VDAC and stimulate its opening. Finally, it is possible that the insertion of Bax into the membrane alters the lipid bilayer structure, resulting in the formation of either lipidic pores or lipid hexagonal phases that would make the membrane permeable to large proteins.

In summary, there are several hypotheses being advanced to explain the mechanism responsible for the permeabilization of the outer mitochondrial membrane. There is experimental evidence arguing for or against each of these mechanisms, which has been related elsewhere [5,10]. The possibility that the mechanisms involved in the permeabilization of the outer mitochondrial membrane may vary according to the death stimulus and the cell type hamper the emergence of a clearer picture of this biological process.

Reactive oxygen species

Reactive oxygen species such as superoxide and the hydroxyl radical, generated by mitochondria, are produced as by-products of normal oxidative metabolism [31]. Superoxide is the product of the addition of an electron to oxygen (Figure 3A). This product can react with iron–sulphur centres containing enzymes leading to the release of iron. Superoxide is removed rapidly by conversion to H_2O_2 in a reaction catalysed by superoxide dismutases. The hydroxyl radical is produced from H_2O_2 in the presence of cupric or ferric ions by the Fenton reaction (Figure 3B). It is a very damaging product that causes peroxidation of proteins, lipids and DNA. A question that has generated a lot of interest is whether oxygen free-radical production is central to the process of apoptosis, is central in some apoptotic responses or is just a side effect. Free radicals have been detected in the early stages of apoptosis caused by p53, TNF or growth factor deprivation. However, recent data reported by Ricci et al. [32] suggest that the production of free radicals

(A)

$$O_2 \xrightarrow{+e^-} O_2^{\cdot -} \xrightarrow[+2H^+]{+e^-} H_2O_2 \xrightarrow{+e^-} OH^\cdot + OH^- \xrightarrow[+2H^+]{+e^-} 2H_2O_2$$

(B)

$$Fe^{2+} + H_2O_2 \longrightarrow Fe^{3+} + OH^- + OH^\cdot$$

$$Cu^+ + H_2O_2 \longrightarrow Cu^{2+} + OH^- + OH^\cdot$$

Figure 3. Generation of oxygen free radicals
(**A**) The sequential addition of electrons to oxygen leads to the formation of anion superoxide, H_2O_2 and the hydroxyl radical. (**B**) Fenton reaction: in the presence of cupric or ferric ions, H_2O_2 can produce the hydroxyl radical.

could occur following permeabilization of the outer mitochondrial membrane and the release of cytochrome c. According to these authors, alteration of complex I of the respiratory chain by caspases is responsible for the generation of free radicals. Blocking caspase activation prevents or delays significantly the generation of free radicals in cells displaying cytochrome c-depleted mitochondria. Although these studies would argue against reactive oxygen species playing a key role in apoptosis, there is no doubt that under certain circumstances, free radicals can also act as signalling molecules and elicit apoptosis. Changes in calcium homoeostasis, or activation of microtubule-associated protein kinases or nuclear factor κB could participate in reactive oxygen species signalling of apoptosis [33].

Conclusion

In conclusion, mitochondria are endosymbiotic organelles of bacterial origin that are involved in the effector phase of apoptosis. Their apoptotic role is regulated by a family of genes which display, at least in part, structural similarity with bacterial toxins. In addition, various components of the death programme, including cytochrome c and AIF, are also found in bacteria. These findings suggest that apoptosis may have evolved together with the endosymbiotic incorporation of aerobic bacteria into ancestral unicellular eukaryotes. Understanding the mechanisms of action of the components of the apoptosis machinery, in particular the relationship with mitochondrial function, is the challenge for the future.

Summary

- *Mitochondria play a role in many apoptotic responses through the release of apoptogenic factors into the cytosol. Production of reactive oxygen species is probably a late event.*
- *However, dysfunctional mitochondria can produce reactive oxygen species that can initiate the apoptotic pathway.*
- *Pro-apoptotic members of the Bcl-2 family enable the permeabilization of the outer mitochondrial membrane and allow the efflux of mitochondrial proteins.*
- *The mechanisms by which pro-apoptotic members of the Bcl-2 family members render the outer mitochondrial membrane leaky remain elusive.*
- *Fragmentation of the mitochondrial network that accompanies permeabilization of the outer mitochondrial membrane plays a role in the release of cytochrome c.*

We thank the Swiss National Science Foundation for their support (grant 3100-061380).

References

1 Harman, D. (1981) The aging process. *Proc. Natl. Acad. Sci. U.S.A.* **78**, 7124–7128

2 Hockenbery, D., Nunez, G., Milliman, C., Schreiber, R.D. & Korsmeyer, S.J. (1990) Bcl-2 is an inner mitochondrial membrane protein that blocks programmed cell death. *Nature (London)* **348**, 334–336

3 Hockenbery, D.M., Oltvai, Z.N., Yin, X.M., Milliman, C.L. & Korsmeyer, S.J. (1993) Bcl-2 functions in an antioxidant pathway to prevent apoptosis. *Cell* **75**, 241–251

4 Hennet, T., Bertoni, G., Richter, C. & Peterhans, E. (1993) Expression of BCL-2 protein enhances the survival of mouse fibrosarcoid cells in tumor necrosis factor-mediated cytotoxicity. *Cancer Res.* **53**, 1456–1460

5 Zamzami, N. & Kroemer, G. (2001) The mitochondrion in apoptosis: how Pandora's box opens. *Nat. Rev. Mol. Cell Biol.* **2**, 67–71

6 Newmeyer, D.D., Farschon, D.M. & Reed, J.C. (1994) Cell-free apoptosis in Xenopus egg extracts: inhibition by Bcl-2 and requirement for an organelle fraction enriched in mitochondria. *Cell* **79**, 353–364

7 Liu, X., Kim, C.N., Yang, J., Jemmerson, R. & Wang, X. (1996) Induction of apoptotic program in cell-free extracts: requirement for dATP and cytochrome *c*. *Cell* **86**, 147–157

8 Susin, S.A., Zamzami, N., Castedo, M., Hirsch, T., Marchetti, P., Macho, A., Daugas, E., Geuskens, M. & Kroemer, G. (1996) Bcl-2 inhibits the mitochondrial release of an apoptogenic protease. *J. Exp. Med.* **184**, 1331–1341

9 Crompton, M. (1999) The mitochondrial permeability transition pore and its role in cell death. *Biochem. J.* **341**, 233–249

10 Desagher, S. & Martinou, J.C. (2000) Mitochondria as the central control point of apoptosis. *Trends Cell Biol.* **10**, 369–377

11 Acehan, D., Jiang, X., Morgan, D.G., Heuser, J.E., Wang, X. & Akey, C.W. (2002) Three-dimensional structure of the apoptosome: implications for assembly, procaspase-9 binding, and activation. *Mol. Cell* **9**, 423–432

12 Adrain, C., Creagh, E.M. & Martin, S.J. (2001) Apoptosis-associated release of Smac/DIABLO from mitochondria requires active caspases and is blocked by Bcl-2. *EMBO J.* **20**, 6627–6636

13 Verhagen, A.M., Coulson, E.J. & Vaux, D.L. (2001) Inhibitor of apoptosis proteins and their relatives: IAPs and other BIRPs. *Genome Biol.* **2**, reviews3009.1–reviews3009.10

14 Suzuki, Y., Imai, Y., Nakayama, H., Takahashi, K., Takio, K. & Takahashi, R. (2001) A serine protease, HtrA2, is released from the mitochondria and interacts with XIAP, inducing cell death. *Mol. Cell* **8**, 613–621

15 Cande, C., Cohen, I., Daugas, E., Ravagnan, L., Larochette, N., Zamzami, N. & Kroemer, G. (2002) Apoptosis-inducing factor (AIF): a novel caspase-independent death effector released from mitochondria. *Biochimie* **84**, 215–222

16 Li, L.Y., Luo, X. & Wang, X. (2001) Endonuclease G is an apoptotic DNase when released from mitochondria. *Nature (London)* **412**, 95–99

17 Wang, X., Yang, C., Chai, J., Shi, Y. & Xue, D. (2002) Mechanisms of AIF-mediated apoptotic DNA degradation in *Caenorhabditis elegans*. *Science* **298**, 1587–1592

18 Borner, C. (2003) The Bcl-2 protein family: sensors and checkpoints for life-or-death decisions. *Mol. Immunol.* **39**, 615–647

19 Conradt, B. & Horvitz, H.R. (1998) The *C. elegans* protein EGL-1 is required for programmed cell death and interacts with the Bcl-2-like protein CED-9. *Cell* **93**, 519–529

20 Yu, J., Wang, Z., Kinzler, K.W., Vogelstein, B. & Zhang, L. (2003) PUMA mediates the apoptotic response to p53 in colorectal cancer cells. *Proc. Natl. Acad. Sci. U.S.A.* **100**, 1931–1936

21 Whitfield, J., Neame, S.J., Paquet, L., Bernard, O. & Ham, J. (2001) Dominant-negative c-Jun promotes neuronal survival by reducing BIM expression and inhibiting mitochondrial cytochrome *c* release. *Neuron* **29**, 629–643

22 Bouillet, P. & Strasser, A. (2002) BH3-only proteins – evolutionarily conserved proapoptotic Bcl-2 family members essential for initiating programmed cell death. *J. Cell Sci.* **115**, 1567–1574

23   Desagher, S., Osen-Sand, A., Montessuit, S., Magnenat, E., Vilbois, F., Hochmann, A., Journot, L.,
      Antonsson, B. & Martinou, J.C. (2001) Phosphorylation of bid by casein kinases I and II regulates
      its cleavage by caspase 8. *Mol. Cell* **8**, 601–611

24   Terradillos, O., Montessuit, S., Huang, D.C. & Martinou, J.C. (2002) Direct addition of BimL to
      mitochondria does not lead to cytochrome c release. *FEBS Lett.* **522**, 29–34

25   Mihara, M., Erster, S., Zaika, A., Petrenko, O., Chittenden, T., Pancoska, P. & Moll, U.M. (2003)
      p53 has a direct apoptogenic role at the mitochondria. *Mol. Cell* **11**, 577–590

26   Sawada, M., Hayes, P. & Matsuyama, S. (2003) Cytoprotective membrane-permeable peptides
      designed from the Bax-binding domain of Ku70. *Nat. Cell Biol.* **5**, 352–357

27   Guo, B., Zhai, D., Cabezas, E., Welsh, K., Nouraini, S., Satterthwait, A.C. & Reed, J.C. (2003)
      Humanin peptide suppresses apoptosis by interfering with Bax activation. *Nature (London)* **423**,
      456–461

28   Suzuki, M., Youle, R.J. & Tjandra, N. (2000) Structure of Bax: coregulation of dimer formation and
      intracellular localization. *Cell* **103**, 645–654

29   Roucou, X., Montessuit, S., Antonsson, B. & Martinou, J.C. (2002) Bax oligomerization in
      mitochondrial membranes requires tBid (caspase-8-cleaved Bid) and a mitochondrial protein.
      Biochem. J. **368**, 915–921

30   Demaurex, N. & Distelhorst, C. (2003) Cell biology: apoptosis – the calcium connection. *Science*
      300, 65–67

31   Raha, S. & Robinson, B.H. (2001) Mitochondria, oxygen free radicals, and apoptosis. *Am. J. Med.*
      Genet. **106**, 62–70

32   Ricci, J.E., Gottlieb, R.A. & Green, D.R. (2003) Caspase-mediated loss of mitochondrial function
      and generation of reactive oxygen species during apoptosis. *J. Cell Biol.* **160**, 65–75

33   Carmody, R.J. & Cotter, T.G. (2001) Signalling apoptosis: a radical approach. *Redox Rep.* **6**, 77–90

5

Death receptors

Harald Wajant

*Department of Molecular Internal Medicine, Medical Polyclinic,
University of Wuerzburg, Roentgenring 11, 97 070 Wuerzburg, Germany*

Abstract

Death receptors {Fas/Apo-1/CD95, TNF-R1 [tumour necrosis factor (TNF) receptor 1], DR3 [death receptor 3], TRAIL-R1 [TNF-related apoptosis-inducing ligand receptor 1], TRAIL-R2, DR6, p75-NGFR [p75-nerve growth factor receptor], EDAR [ectodermal dysplasia receptor]} form a subgroup of the TNF-R superfamily that can induce apoptosis (programmed cell death) via a conserved cytoplasmic signalling module termed the death domain. Although death receptors have been recognized mainly as apoptosis inducers, there is growing evidence that these receptors also fulfil a variety of non-apoptotic functions. This review is focused on the molecular mechanisms of apoptotic and non-apoptotic death receptor signalling in light of the phenotype of mice deficient in the various death receptors.

Introduction

The cytokines of the TNF (tumour necrosis factor) ligand family have been appreciated mainly as regulators of the immune system, but there is also a growing number of examples implicating these proteins in developmental processes [1,2]. The ligands of the TNF family are organized as homotrimers and are mostly expressed as type II membrane proteins. In addition, soluble forms can be derived from the membrane-bound ligands by proteolytical processing or by alternative splicing. The members of the TNF ligand family

[1]E-mail *harald.wajant@mail.uni-wuerzburg.de*

exert their functions through interaction with a complementary family of type I membrane proteins – the members of the TNF-R (TNF receptor) superfamily [1–3]. Based on crystallographic studies and cross-linking experiments, it is commonly assumed that a ligand trimer binds three molecules of its corresponding receptor at the interfaces of neighbouring ligand subunits [3]. Remarkably, a ligand trimer does not interact sequentially with receptor monomers, but rather binds and reorganizes signalling-incompetent complexes of receptors, which are pre-assembled via an N-terminal domain – the PLAD (pre-ligand-binding assembly domain) [4]. Some receptors, such as CD95, TRAIL-R2 (TNF-related apoptosis-inducing ligand receptor 2), TNF-R2 and CD40, can be activated by both the soluble and the membrane-bound form of their corresponding ligand with comparable efficiency, whereas, in other cases, appropriate receptor activation is only possible by the membrane-bound form. Soluble ligands that do not stimulate their corresponding receptors nevertheless bind to these receptors. Thus, in the TNF ligand and receptor families, ligand binding is not in each case sufficient for receptor activation.

Members of the TNF-R superfamily are characterized by the presence of up to six copies of a cysteine-rich domain in their extracellular part. With respect to function and structure of the intracellular domain, some members of the TNF-R superfamily can be grouped into the death receptor group (Figure 1). The hallmark of these receptors is a cytoplasmic protein–protein-interaction domain called the death domain, which is critically involved in apoptosis induction by these receptors [5]. Death receptors interact with death-domain-containing adaptor proteins by homophilic association of their death domains. The essential role of the death-domain of death receptors for initiation of apoptosis is based on its function as a docking site for the recruitment and activation of cytoplasmic death-domain-containing proteins which trigger the apoptotic response (for details, see below). Although the death domain was originally identified in the context of apoptosis induction, it is now clear that death-domain-mediated protein–protein interactions also transduce non-apoptotic functions of death receptors (Table 1). Stimulation of non-death-domain-containing members of the TNF-R superfamily (e.g. CD40, TNF-R2) can also result in apoptosis induction by transcriptional up-regulation of ligands of death receptors [6]. Thus non-death-domain-containing receptors induce apoptosis via death receptors, too.

CD95 and the TRAIL death receptors

Mechanisms of apoptosis induction
CD95 (Apo-1, Fas) has been originally identified as a cell-surface antigen targeted by apoptosis-inducing antibodies [7,8]. Molecular cloning identified CD95 as a typical member of the TNF-R superfamily and deletion

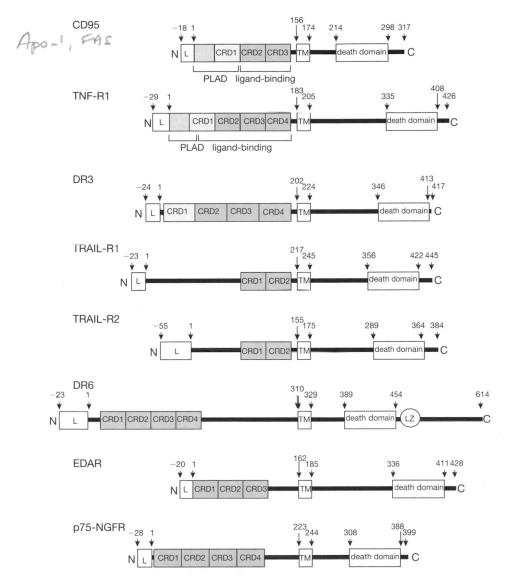

Figure 1. Domain architecture of death receptors
Numbering is based on the mature proteins. CRD, cysteine-rich domain; L, leader peptide; LZ, leucine zipper; TM, transmembrane domain. For other abbreviations see the text.

mutagenesis of CD95 led to the definition of the death domain. CD95 is activated by membrane-bound CD95L, but not by soluble CD95L, which can be released from the membrane-bound form by metalloproteases [9]. TRAIL has been identified in an EST (expressed sequence tag) database screen for novel members of the TNF ligand family. TRAIL interacts with five members of the TNF-R superfamily, namely TRAIL-R1 and TRAIL-R2, both containing a death domain, TRAIL-R3, which is attached to the plasma

Table 1. *In vivo* functions of death receptors

DR, death receptor; p75-NGFR, p75-nerve growth factor receptor; EDAR, ectodermal dysplasia receptor; TCR, T-cell receptor.

Receptor	Apoptosis-related function	Non-apoptotic function	Reference
TNF-R1	–	Endothelial activation	[17]
		Organogenesis of lymphoid organs	
		Regulation of osteoclastogenesis	
		Regulation of adipocyte metabolism	
		Defence from intracellular pathogens	
Fas	Tumour surveillance	Liver regeneration	[9]
	Immune privilege	Positive selection of thymocytes	
	Activation-induced cell death	Chemokine induction	
	T- and B-cell homoeostasis		
TRAIL-R1	Tumour surveillance	–	[10]
TRAIL-R2	Tumour surveillance	–	[10]
DR3	Negative selection of thymocytes	–	[20]
	TCR-induced cell death		
DR6	–	Inhibition of B-cell proliferation	[21,22]
p75-NGFR	Neuronal cell death	Differentiation and repair of neuronal cells	[23]
EDAR	–	Development of hair, teeth and sweat glands	[24]

membrane via a glycosylphosphatidylinositol anchor, and TRAIL-R4, which contains a truncated death domain devoid of apoptosis-inducing capabilities. TRAIL also interacts with osteoprotegerin (the fifth TRAIL receptor), a soluble decoy receptor, which, in addition, binds to RANK [receptor activator of NF-κB (nuclear factor κB)] ligand, a TNF family member involved in the regulation of osteoclastogenesis [10].

CD95L-induced apoptosis is an effector mechanism of cytotoxic T-cells and has been implicated in activation-induced cell death of CD8$^+$ T-cells. CD95L, CD95 and the TRAIL death receptors are induced by the tumour suppressor p53, suggesting an involvement of these molecules in p53-induced apoptosis and tumour surveillance [9,10]. Like CD95, TRAIL-R2 is only efficiently activated by the membrane-bound form of its cognate ligand, whereas TRAIL-R1 is triggered both by soluble and membrane-bound TRAIL. CD95 and the TRAIL death receptors activate the apoptotic machinery by a common mechanism. The initial event for both apoptotic and non-apoptotic signalling via these receptors, is ligand-mediated reorganization of the PLAD-assembled receptor complexes into a state which facilitates the assembly of the so-called DISC (death-inducing signalling complex). The PLAD does not overlap with the extracellular parts of the receptor responsible for ligand binding. The functional importance of the preformed CD95 complexes is evident from the fact that deletion of the PLAD completely abrogates CD95 signalling. In addition to ligand and receptor, the DISC contains the FADD (Fas-associated death domain) protein and pro-caspase-8. FADD, which is an adaptor protein, binds via its C-terminal death domain to ligated CD95, TRAIL-R1 or TRAIL-R2, and pro-caspase-8 is recruited into the DISC by binding to the N-terminal death-effector domain of FADD via its own N-terminal death-effector domains (Figure 2). As the death domain, death-effector domains mediate protein–protein interactions by homophilic association. Remarkably, death domains, as well as death-effector domains, consist of a conserved structural arrangement of eight α-helices, the death-domain fold [11].

The DISC contains at least three molecules of one of the death receptors CD95, TRAIL-R1 or TRAIL-R2, but most likely more, as there is evidence that trimeric ligand–receptor complexes secondarily aggregate into supramolecular clusters. Thus, as a consequence of ligand-induced DISC assembly, several pro-caspase-8 molecules are brought together in close proximity resulting in their activation by dimerization (Figures 3a–3c). Subsequently, the mature caspase-8 heterotetramer, which is also stable and active upon release from the DISC, is generated by autoproteolytic processing into the p18 and p10 subunits (Figures 3d and 3e) [12,13]. A first cleavage event occurs between the p18 and p10 subunits of the caspase homology domain of pro-caspase-8, resulting in a FADD-bound p43/41 intermediate (Figure 3d), which contains the death-effector domains and the p18 subunit. Notably, the p10 subunit remains non-covalently associated with the p43/41 intermediate. A second cleavage event, which releases the mature enzymically active p18/p10 heteromer of caspase-8

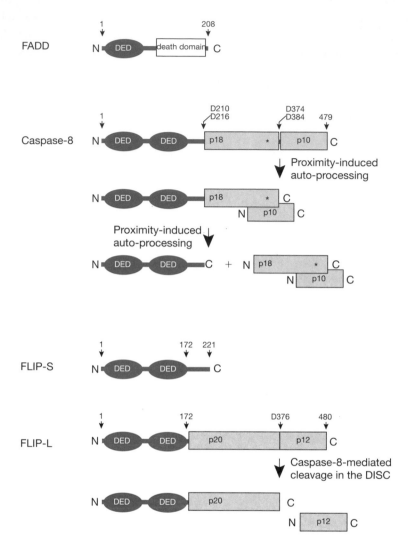

Figure 2. Domain architecture and processing of FADD, caspase-8 and FLIP
p18 and p10, and p20 and p12 represent the subdomains of the caspase homology domain of caspase-8 and FLIP, respectively. Please note that only caspase-8 has a functional active site (*). DED, death-effector domain.

into the cytoplasm, occurs between the death-effector domains and the p18 subunit. After replacement of the FADD-bound death-effector domains of caspase-8 by another pro-caspase-8 molecule, the mature p18/p10 hetero-tetramer is released from the DISC and a new cycle of caspase-8 processing occurs (Figure 3). Thus the DISC acts as a pro-caspase-8-converting complex.

Downstream of caspase-8 activation, apoptosis induction takes place in a cell-type specific manner. In some cell types, designated as type I cells, caspase-8 processes and activates effector caspases, especially caspase-3, directly with an efficiency sufficient to ensure apoptotic cell death. However, in anoth-

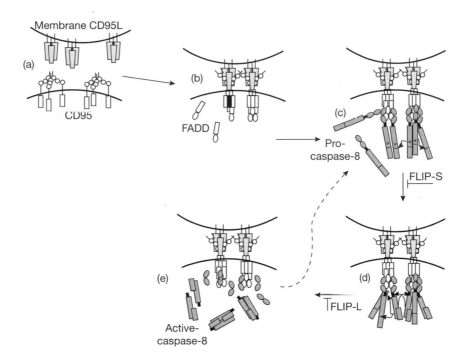

Figure 3. Membrane CD95L induces formation of the DISC, a caspase-8-converting receptor complex

Interaction of membrane CD95L and preformed CD95 complexes (**a**) leads to reorganization of these complexes, binding of FADD (**b**) and FADD-mediated recruitment of several pro-caspase-8 molecules into the CD95L–CD95 complex (**c**). As a consequence of the locally increased concentration of pro-caspase-8 in the DISC, pro-caspase-8 transiently assembles into DISC-bound active dimers (labelled *), which convert to the stable mature form of caspase-8 in two steps by auto-processing (**d, e**).

er type of cells (type II cells) apoptosis induction by DISC-mediated caspase-8 activation is dependent on an amplification loop, which is triggered by caspase-8-mediated cleavage of Bid, a BH3 domain-only member of the Bcl-2 protein family. The resulting cleavage product (truncated Bid, tBid) can induce Bax and/or Bak-mediated release of cytochrome *c* and other pro-apoptotic mediators, including the second mitochondria-derived activator of caspase/direct IAP (inhibitor of apoptosis protein)-binding protein [Smac (second mitochondrial activator of caspases)/Diablo] and the serine protease HtrA2/Omi, from mitochondria. Seven molecules of ATP, cytosolic cytochrome *c* and the scaffold protein apoptotic protease-activating factor 1 (Apaf-1) assemble to form a complex called the apoptosome [14], which is able to activate caspase-9. Smac/Diablo and HtrA2/Omi interfere with the caspase-inhibitory function of members of the IAP family, thereby promoting apoptosis. Caspase-9 activates the effector caspase-3, thus enhancing the effect of initially DISC-activated caspase-8. Remarkably, active caspase-3 itself is able to process its upstream activators, caspase-8 and caspase-9, establishing a self-amplifying loop of caspase activation (Figure 4). Experimentally, type I and type II cells can be distinguished by the effect of the anti-apoptotic Bcl-2 protein. Whereas Bcl-2 expression has no effect on death receptor-induced apoptosis in type I cells, Bcl-2 confers protection in type II cells by interference with Bax/Bak-mediated release of the above-mentioned pro-apoptotic mitochondrial factors. Noteworthy, death receptors can also trigger cell death by necrosis, which occurs independently from, or is even blocked by, caspases. Death-receptor-induced necrosis bifurcates at the level of FADD from apoptosis induction. Thus, whereas caspase-8 is dispensable for necrosis induction, FADD together with the death-domain-containing kinase RIP (receptor-interacting protein) has an essential role in this process [15].

A couple of additional proteins (see Table 2) that facilitate apoptosis induction have also been identified as part of the DISC of CD95, TRAIL-R1 and TRAIL-R2. However, the precise functions and relevance for apoptosis induction of these DISC components are poorly understood. Other proteins that can be recruited to CD95 or the TRAIL death receptors interfere with DISC action or have been implicated in the activation of non-apoptotic sig-

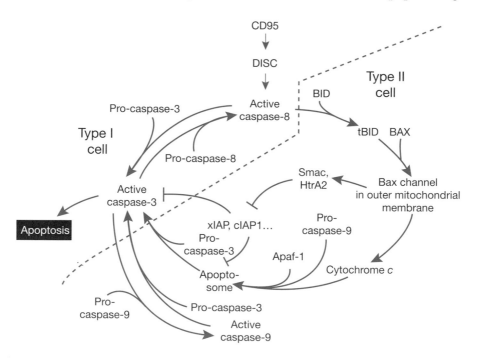

Figure 4. Two CD95 signalling pathways lead to apoptosis
In type I cells, the CD95 DISC triggers strong caspase-8 activation leading to activation of other caspases, such as caspase-3, and apoptosis. In type II cells, only a sub-optimal amount of active caspase-8 is generated by the DISC, making apoptosis induction dependent on a mitochondrial amplification loop. As active caspase-3 is able to process and activate pro-caspase-8 and pro-caspase-9 outside the DISC and the apoptosome, respectively; the activity of these initiator caspases is enhanced by positive-feedback loops. While FLIP-S blocks the first cleavage step, FLIP-L interferes with the second step. cIAP, cellular inhibitor of apoptosis; xIAP, X-linked inhibitor of apoptosis.

Table 2. Proteins interacting directly with death receptors

Abbreviations used: DR, death receptor; p75-NGFR, p75-nerve growth factor receptor; DAP1/3, death-associated protein 1/3; EDAR, ectodermal dysplasia receptor; EDARDD, EDAR-associated death domain protein; FAF-1, Fas-associated factor; FAP-1, Fas-associated phosphatase-1; IRAK1, interleukin-1-receptor-associated kinase 1; LFG, lifeguard; MADD, mitogen-activated-protein-kinase-activating death-domain protein; NADE, p75-NTR-associated cell-death executor; NRAGE, neurotrophin receptor-interacting MAGE homologue; NRIF, neurotrophin-receptor-interacting factor; PKCζ, atypical protein kinase Cζ; SHP-1; Src homology domain 2 (SH2)-containing tyrosine phosphatase-1; SODD, silencer of death domain; TRADD, TNF-R1-associated death-domain protein; TRAF6, TNF-R2-associated factor 6; Ubc9, ubiquitin-conjugating enzyme 9; JNK, c-Jun N-terminal kinase; GM-CSF, granulocyte/macrophage colony-stimulating factor; MAGE, melanoma-associated antigen; p75-NTR, p75 neurtophil receptor; FAN, factor associated with neutral SMase activation; DAXX, death-associated protein; FLASH, Flice-associated huge protein.

Receptor	Associated protein	Relevance for death receptor signalling	Reference
TNF-R1	DAP1	Poorly defined	[17]
	FAN	Recruitment of neutral SMase	
	MADD	Mitogen-activated protein kinase signalling	
	Sentrin	Poorly defined	
	SODD	Inhibition of auto-aggregation	
	TRADD	Recruitment of FADD and caspase-8, apoptosis induction	
		Recruitment of TRAF2, NF-κB and JNK signalling	
		Recruitment of RIP, NF-κB signalling	
	Ubc9	Poorly defined	
Fas	DAXX	JNK signalling	[9]
	Ezrin	Poorly defined	
	FADD	Recruitment of caspase-8, apoptosis induction	
	FAP-1	Inhibition of Fas translocation to the cell surface	
	FAF-11	Poorly defined	
	FLASH	Implicated in DISC formation	
	p100	Implicated in DISC formation	
	RIP	Mediates Fas-induced necrosis	
	LFG	Poorly defined	
	Sentrin	Poorly defined	
	SHP1	Mediates cross-talk with GM-CSF signalling in a cell-type-specific manner (neutrophils)	
TRAIL-R1	DAP3	Implicated in DISC formation	[10]
	FADD	Recruitment of caspase-8, apoptosis induction	
	SHP1	Mediates cross-talk with GM-CSF signalling in a cell-type-specific manner (neutrophils)	
TRAIL-R2	DAP3	Implicated in DISC formation	[10]
	FADD	Recruitment of caspase-8, apoptosis induction	
	SHP1	Mediates cross-talk with GM-CSF signalling in a cell-type-specific manner (neutrophils)	

contd ☞

Table 2. contd

Receptor	Associated protein	Relevance for death receptor signalling	Reference
DR3	TRADD	Recruitment of FADD and caspase-8, apoptosis induction	[20]
		Recruitment of TRAF2, NF-κB signalling	
DR6	–	–	–
p75-NGFR	SC1	Poorly defined	[23]
	NRIF1	Implicated in apoptosis induction	
	NRIF2	Implicated in apoptosis induction	
	NADE	Implicated in apoptosis induction	
	NRAGE	Implicated in JNK-mediated apoptosis induction	
	Necdin	Poorly defined, related to NRAGE	
	MAGE-H1	Poorly defined, related to NRAGE	
	RIP2	NF-κB activation	
	IRAK1	NF-κB activation	
	p62	NF-κB activation	
	TRAF6	NF-κB activation	
	PKCζ	NF-κB activation	
EDAR	EDARDD	Recruitment of TRAF2, NF-κB signalling	[24]

nalling pathways. The most important negative regulator of death receptor-induced apoptosis seems to be the FLIP (Flice-like inhibitory protein), a caspase-8 homologue devoid of proteolytic activity that is expressed in several isoforms [16]. So far, FLIP-short (FLIP-S) and FLIP-long (FLIP-L) are the best studied of these isoforms. FLIP-L consists of two N-terminal death-effector domains and a caspase-homology domain without catalytic activity. In its overall architecture, FLIP-L resembles caspase-8, whereas FLIP-S comprises only the two N-terminal death-effector domains of FLIP-L (Figure 2). Both FLIP splice forms can be incorporated into the DISC of death receptors and block apoptosis induction. However, DISC-induced maturation of caspase-8 is inhibited at different steps. While FLIP-S completely blocks processing and activation of pro-caspase-8, FLIP-L forms heterodimers with pro-caspase-8, which have the ability to catalyse cleavage of pro-caspase-8 into p43/41 and p10. This partially processed caspase form and also the heteromer of FLIP-L and pro-caspase-8 are proteolytically active, but are not sufficient for apoptosis induction [17] because they are not released from the DISC. It is tempting to speculate that mature caspase-8 released from the DISC is necessary for apoptosis induction whereas DISC-bound FLIP-L pro-caspase-8 heteromers act on DISC-localized substrates. This may also explain, at least partially, the differential regulation of non-apoptotic CD95 signalling by FLIP-S and FLIP-L. For example, up-regulation of the proto-oncogene c-*fos* by CD95L and TRAIL is enhanced in FLIP-S-expressing cells, but it is completely blocked in the same type of cell when FLIP-L is overexpressed [18]. Furthermore, FLIP-L, but not FLIP-S, favours CD95-mediated activation of the transcription factor NF-κB. Thus the isoforms of FLIP do not simply inhibit apoptosis induction by death receptors, but rather define the overall quality of cellular response towards stimulation of death receptors.

The prominent apoptosis-inducing features of CD95 and the TRAIL death receptors are mirrored in the phenotypes of mice deficient in or with defective receptors or ligands. The mouse strains lpr (for lymphoproliferation) and gld (for generalized lymphoproliferative disease) have defects in CD95 and CD95L and develop lymphoadenopathy, splenomegaly and display increased levels of auto-antibodies [19], most likely due to the absence of CD95/CD95L-mediated activation-induced cell death of activated T-cells. Furthermore, lpr mice display enhanced formation of metastases in some experimental tumour models, arguing for a role of the CD95/CD95 system in tumour surveillance. Increased tumour initiation and metastasis have also been reported for TRAIL-deficient mice [20,21]. These data correspond with the reported susceptibility of many tumour cells for death-receptor-induced apoptosis. Furthermore, reduced negative selection of thymocytes has been reported in TRAIL-deficient mice [22].

Non-apoptotic signalling of CD95 and the TRAIL death receptors

There is growing evidence that CD95 and the TRAIL death receptors can also transduce proliferating and/or activating signals [9,10]. Activation of CD95 is sufficient to drive proliferation of human diploid fibroblasts and significantly enhances TCR (T-cell receptor)-triggered proliferation of thymocytes and T-cells. Moreover, there is evidence for a pro-proliferative role of CD95 in the regenerative liver response after partial hepatectomy. Although CD95-mediated activation of extracellular-signal-regulated kinase (ERK)-1/2 and up-regulation of c-*fos* have been demonstrated in fibroblasts and T-cells, the signalling pathways leading to these hallmarks of proliferation are poorly defined. Remarkably, there is evidence for an involvement of non-apoptotic CD95-mediated caspase activation in the early phase of T-cell proliferation. In particular, this occurs upon sub-optimal stimulation of the TCR. Moreover, the death-domain-containing adaptor protein FADD has also been implicated in T-cell proliferation. As T-cell activation results in strong up-regulation of CD95L, a model of T-cell activation seems feasible, in which a TCR-induced autocrine loop of CD95 activation, and maybe activation of other death receptors, stimulates a non-apoptotic caspase-mediated pathway contributing to T-cell proliferation. However, the components of this pathway and its downstream targets are largely unknown as yet. At a first glance, the phenotype of mice with genetic defects of CD95 or CD95L, both showing a dominant auto-immune disease, is contradictory to a role of CD95 activation in T-cell proliferation. However, this could be due to redundant action of other death receptors or unknown FADD-dependent modes of caspase activation. There is also some evidence arguing for a differential importance of CD95 stimulation for T-cell proliferation, dependent on the subtype of T-cell and the quality of the initial TCR signal. Moreover, impaired CD95 function results in drastically reduced numbers of thymocytes due to reduced positive selection. An unexpected role of CD95 and CD95L in inflammation has been revealed by studies that have investigated the use of CD95L expression to protect transplants from rejection by the immune system [9]. Based on the concept that CD95L expression in allografts could induce apoptosis in infiltrating inflammatory cells, it was expected that CD95L-positive transplants exert reduced rejection. Unexpectedly, in most cases, CD95L expression failed to confer protection of transplants and often even enhanced the development of granulocytic/neutrophilic infiltrates. While some studies attributed the latter effect to a direct chemotactic function of CD95L, other studies could not confirm this and suggested that neutrophil-attracting chemokines, e.g. interleukin 8 or monocyte chemotactic protein-1, are released by apoptotic cells or induced by CD95L in neighbouring cells. Indeed, CD95-induced chemokine production has been described in several cell types. In some cases, this may be associated with apoptosis induction and caspase activation, but may also occur independently from apoptosis. Although in some studies, activation of NF-κB and JNK (c-Jun N-terminal kinase), both important inducers of interleukin 8 and monocyte chemotactic protein-1, correlates with CD95-mediated chemokine production, the relevance of these pathways for CD95-mediated inflammation *in vivo* is still poorly investigated. Future studies have to define the molecular architecture of the signalling pathways involved in CD95-mediated inflammation, and in particular have to clarify whether these pathways work independently or act together in an integrated way.

TNF-RI

TNF-RI-induced apoptosis

Like CD95 and the TRAIL death receptors, TNF-R1 induces apoptosis by a FADD- and caspase-8 dependent pathway [23]. These receptors are also similar with respect to their ability to activate the NF-κB pathway. Nevertheless, in contrast to CD95 and the TRAIL death receptors, which show prominent apoptosis induction *in vivo*, the *in vivo* functions of TNF-R1 are clearly dominated by its non-apoptotic NF-κB-related gene-inducing capabilities (Table 1). This principal difference between TNF-R1 and CD95 and the TRAIL death receptors is most likely caused by the way in which these receptors are linked to FADD. As described above, CD95 and the TRAIL death receptors can directly interact with FADD. However, TNF-R1 and FADD do not directly associate with each other. Rather, recruitment of FADD into the TNF-R1 signalling complex depends on an additional death-

domain-containing adaptor protein, called TRADD (TNF-R1-associated
death-domain protein). Importantly, TRADD not only links TNF-R1 to
FADD and caspase-8 via its C-terminal death domain, but also mediates
recruitment of the TRAF2 (TNF-R2-associated factor 2) adaptor protein
(Figure 5). The latter interferes with TNF-R1-induced apoptosis by two
interrelated mechanisms [23]. First, TRAF2, possibly together with TRAF1,
recruits the cIAPs (cellular IAPs) 1 and 2 into the TNF-R1 signalling complex,
leading to the inhibition of caspase-8 processing (Figure 5). Secondly, TRAF2
is involved in TNF-R1-induced activation of the NF-κB pathway, which
regulates a variety of anti-apoptotic factors, including some of the IAP
proteins, TRAF1 and cellular FLIP (Figure 5). TRADD and the TRAF2–IAP
complex are major components of the TNF-R1 signalling complex, but do not,
or only modestly, bind to CD95 and the TRAIL death receptors. Thus in the
absence of FLIP, TRADD-mediated recruitment of the anti-apoptotic

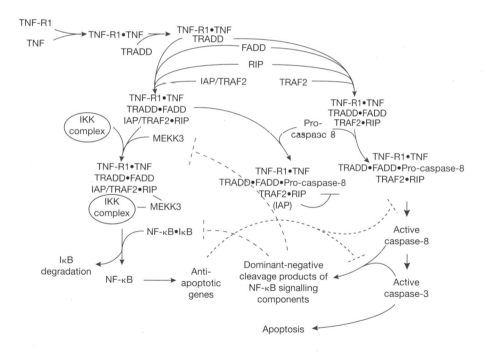

Figure 5. TNF-R1 signalling pathways leading to NF-κB and apoptosis induction
Progressive model of TNF-induced NF-κB activation and apoptosis induction. Upon ligand-
induced reorganization of PLAD-assembled TNF-R1 complexes, the death-domain-containing
adapter protein TRADD can interact with the receptors and serves as an assembly platform for
the recruitment of NF-κB-inducing (TRAF2, RIP) and apoptosis-inducing (FADD, caspase-8) proteins. The accessibility of anti-apoptotic IAP proteins, that can be secondarily recruited to TNF-
R1 via TRAF2, determines the final outcome of TNF stimulation. Proteins acting together in a
complex are connected by black circles. Noteworthy, apoptosis induction and NF-κB activation
are inhibitory to each other. Thus NF-κB-induced proteins block apoptosis induction and activated caspases cleave components of the NF-κB signalling pathway into fragments acting in a dominant-negative manner on their non-cleaved counterparts. MEKK1, mitogen-activated protein
kinase/extracellular signal-regulated kinase kinase kinase 1.

TRAF2–IAP complex prevents caspase-8 activation by TNF-R1, but not by CD95 and the TRAIL death receptors. Directly after TNF-R1 stimulation, blockage of TNF-R1-mediated caspase-8 activation exclusively depends on immediately available, therefore pre-existing, anti-apoptotic factors. However, at later time points, TNF-R1-induced NF-κB activation contributes to inhibition of apoptosis induction by the TRAF2–IAP complex by up-regulation of cIAP1 and cIAP2, and even confer a broader anti-apoptotic status by up-regulation of anti-apoptotic proteins, such as FLIP and Bfl1, that interfere with apoptosis induction by a wider range of stimuli. Noteworthy, ongoing apoptosis can block the NF-κB signalling pathway by caspase-mediated cleavage of some of its components (Figure 5). The dual anti-apoptotic role of TRAF2 in TNF-R1 signalling described above also explains the selective enhancement of TNF-R1-induced apoptosis by TNF-R2, a non-death-domain-containing member of the TNF-receptor superfamily. TNF-R2 directly binds TRAF2 and thus reduces the pool of freely available cytoplasmic TRAF2–IAP complexes within a few hours upon stimulation by efficient recruitment of these proteins and their subsequent proteasomal degradation. Thus co-stimulation of both TNF-Rs leads to TNF-R2-dependent depletion of TRAF2–IAP complexes and enhancement of TNF-R1-induced, but not CD95- or TRAIL death-receptor-induced, apoptosis.

Non-apoptotic TNF-R1 signalling

The most important non-apoptotic cellular response of TNF-R1 stimulation is the activation of transcription factors of the NF-κB family. These transcription factors act as homo- or hetero-dimers of NF-κB proteins [24]. They are normally sequestered in the cytoplasm in an inactive form by interaction with IκB (inhibitory κB) proteins or by a C-terminal domain found in some members of the NF-κB family. Signal-induced activation of NF-κB transcription factors requires at least two events. First, translocation into the nucleus and, secondly, stimulation of the transactivation activity of the NF-κB proteins by phosphorylation. Translocation of NF-κB proteins is triggered by phosphorylation- and ubiquitination-dependent proteolytic degradation of associated IκB proteins or by proteolytic removal of the auto-inhibitory C-terminal domain of NF-κB precursors. Degradation of IκB proteins, as well as proteolytic processing of NF-κB precursors, is mediated by the proteasome. TNF, like other stimuli, triggers IκB phosphorylation and degradation by activation of the IKK (IκB kinase) complex, which consists of the regulatory subunit NEMO (NK-κB essential modulator)/IKKγ, the chaperones heat-shock protein 90 and Cdc37 and of two related kinases, IKK1 and IKK2. Upon TNF stimulation TRAF2 recruits the IKK complex into the TNF-R1 signalling complex (Figure 5). The RIP, another component of the TNF-R1 signalling complex, and MEKK3 then activate, by a poorly understood mechanism, IKK1 and IKK2 by phosphorylation of serine residues in the activation loop of these kinases. Maybe this function is redundantly fulfilled by the many known TRAF2-associated proteins [25] and can therefore differ from cell to cell.

DR3 (death receptor 3)

DR3 was identified more than 5 years ago by several groups as a death receptor displaying the highest homologies with TNF-R1, with predominant expression in lymphocytes. The ligand of DR3 is TL1A, a recently discovered member of the TNF ligand family. Earlier studies claiming that TWEAK, another member of the TNF family, is a ligand of DR3 were not reproducible. Generation of DR3-deficient mice revealed a role of this receptor in negative selection and anti-CD3-induced cell death of T-cells [26]. So far, it appears that DR3 elicits the same cellular pathways as TNF-R1, using the same set of receptor-associated proteins. Thus DR3 induces apoptosis by TRADD-mediated recruitment of FADD and caspase-8, and triggers NF-κB activation via RIP and TRAF2. The non-redundant functions of DR3 and TNF-R1 are therefore most likely to be related to the different expression patterns of these receptors.

DR6

DR6 is highly conserved and was originally identified in an EST database search for novel members of the TNF-R superfamily. Comparative sequence analyses indicated that DR6 contains a death domain, which is unusually located adjacent to the transmembrane domain, but not at the C-terminal end of the molecule as in other death receptors. Preliminary over-expression studies showed that DR6 interacts weakly with TRADD, but not with FADD, via its death domain. While DR6 robustly activates the JNK pathway independently of its death domain, only a moderate capacity of the receptor to induce NF-κB or apoptosis upon over-expression has been reported. Mice with targeted disruption of DR6 show increased cell expansion, and enhanced survival and humoral responses of activated B-cells [27,28]. They also display enhanced CD4$^+$ T-cell proliferation and increased Th2 cytokine production. This phenotype correlates with an increase in nuclear cRel, a member of the NF-κB family, and nuclear factor of activated T-cells c (NF-ATc) in B- and T-cells, of DR6-deficient mice [24]. There is also reduced JNK activity in activated CD4$^+$ T-cells of DR6$^{-/-}$ mice. This is in good accordance with the increase of nuclear NF-ATc in these cells, because JNK has an inhibitory effect on calcineurin-mediated activation of NF-ATc. Although enhanced B-cell proliferation in DR6-deficient mice is partly associated with reduced apoptosis, this effect reflects increased activity of cRel, rather than direct DR6-mediated triggering of apoptosis. Thus together, these data suggest that DR6 fulfills its immune-regulatory functions largely independently from direct apoptosis induction.

p75-NGFR (p75-nerve growth factor receptor)

p75-NGFR is the founding member of the TNF-R superfamily and is involved in the regulation of survival, differentiation and repair of neuronal cells [29]. p75-NGFR is unique in the TNF-R superfamily with respect to two aspects. First, neurotrophins [e.g. NGF (nerve growth factor) and BNDF (brain-derived neurotrophic factor)], the ligands of p75-NGFR, do not belong to the TNF ligand family. Secondly, p75-NGFR co-operates with the unrelated receptor tyrosine kinase TrkA in the formation of high-affinity binding sites for NGF, whereas the other members of the TNF-R family do not form heteromeric complexes. Nevertheless, both receptors are able to signal in the absence of the other receptor. The number of helices and the overall fold of the death domain of p75-NGFR are largely similar to other death domains, but, in contrast to other death domains, the death domain of p75-NGFR neither self-associates nor interacts with other death domains. In the absence of TrkA signalling, NGF induces apoptosis via p75-NGFR in hippocampal neurons and oligodendrocytes independently from FADD and caspase-8. Indeed, in many cell types, p75-NGFR apparently signals apoptosis induction by a unique cytoplasmic juxtamembrane domain and not via its death domain. Moreover, in these cells, p75-NGFR-mediated apoptosis is dependent on JNK pathway-induced transcription/translation. Taken together, these data suggest that p75-NGFR-mediated apoptosis is primarily initiated by transcriptional up-regulation of pro-apoptotic factors rather than by direct engagement of apoptotic caspases. Nevertheless, there is some evidence that p75-NGFR also signals apoptosis via its death domain under defined circumstances [29]. More than a dozen proteins have been identified as part of the p75-NGFR signalling complex (Table 2) and four of them have been implicated in apoptosis induction, namely SC-1, neurotrophin receptor-interacting factor (NRIF), p75NTR-associated cell death executor (NADE) and neurotrophin receptor-interacting MAGE homologue (NRAGE). While SC-1 and NRIF bind to the cytoplasmic juxtamembrane part of p75-NGFR, NADE associates with the C-terminal death domain. However, how these molecules are linked to apoptosis induction and whether JNK activation is involved in this respect has not been resolved so far. For NRAGE, a role in both, p75-NGFR-mediated JNK activation and apoptosis induction, has been found, but the part of p75-NGFR that associates with this protein has not yet been mapped. Most of the other p75-NGFR-associated proteins [RIP2, IRAK (interleukin-1 receptor-associated kinase 1), p62, TRAF6, PKCζ (atypical protein kinase Cζ)] have been implicated in p75-NGFR-mediated NF-κB activation, but their molecular mode of action is still elusive. As NF-κB activation and TrkA signalling can modulate apoptosis, p75-NGFR-induced apoptosis is most likely to be highly dependent on cell type, differentiation state and experimental circumstances.

EDAR (ectodermal dysplasia receptor)

The death receptor EDAR and its ligand, the EDA-A1 isoform of ectodysplasin, have a pivotal role in the development of hair, teeth and sweat glands [30]. Mutations in these proteins are the molecular cause of the name-giving disorder hypohidrotic/anhidrotic ectodermal dysplasia in males and the related phenotype in *downless* and *tabby* mice [30]. Ectodermal dysplasia is characterized by defects in hair-follicle induction, lack of sweat glands and malformation of teeth. Mice with a defect in the EDAR-associated death domain protein (EDARDD), a death-domain-containing adaptor protein that associates with the EDAR death-domain, show these typical ectodermal dysplasia symptoms too. The same applies for mice deficient in NEMO/IKKγ or TRAF6, both involved in NF-κB signalling. Together these data suggest that ectodermal dysplasia is most likely caused by the lack of EDAR-mediated NF-κB activation. Thus the *in vivo* functions of EDAR are mainly independent from apoptosis induction. In accordance with these genetic data, EDAR is a strong inducer of NF-κB activation, but neither stimulates caspase activity nor interacts with TRADD or FADD.

Conclusion

The death domain is a structurally conserved protein–protein-interaction module. Death-domain-containing receptors and adaptor proteins have been identified *in vitro* and *in vivo* as essential components of receptor-induced apoptosis. However, recent research has shown that death receptors also initiate a plethora of non-apoptotic cellular responses. Indeed, some death receptors seem completely devoid of pro-apoptotic capabilities. Remarkably, one of the most important non-apoptotic cellular responses induced by death receptors is activation of the anti-apoptotic NF-κB pathway. It will be central to the understanding of death-receptor biology to learn how the balance between apoptosis induction and NF-κB activation upon death-receptor triggering is regulated at the molecular level.

Summary

- *The death domain is a structurally conserved protein–protein-interaction module.*
- *Death receptors (CD95, TNF-R1, DR3, TRAIL-R1, TRAIL-R2, DR6, p75-NGFR, EDAR) form a subgroup of the TNF-R superfamily containing a death domain.*
- *Some, but not all, death receptors induce apoptosis by direct or indirect recruitment of the death-domain-containing adaptor protein FADD and caspase-8.*
- *The enzymically inactive caspase-8 homologue FLIP is a major regulator of death-receptor-induced apoptosis.*

- *TNF-R1 signals NF-κB activation by TRAF2-mediated recruitment of the IKK complex and RIP/MEKK3-mediated activation of its kinases.*
- *The NF-κB pathway up-regulates a variety of anti-apoptotic factors.*
- *The NF-κB pathway is blocked by caspase-mediated cleavage of some of its components during apoptosis.*
- *TNF-R1-induced, but not CD95- and TRAIL death-receptor-induced, apoptosis is regulated by a TRAF2–IAP complex.*
- *p75-NGFR-induced apoptosis typically occurs indirectly via the JNK pathway.*
- *All death receptors can also engage non-apoptotic pathways.*

This work was supported by the Deutsche Forschungsgemeinschaft (grant Wa 1025/11-1 Sonderforschungsbereich 495 project A5) and Dr Mildred Scheel Stiftung für Krebsforschung grant 10-1751.

References

1. Zhou, T., Mountz, J.D. & Kimberly, R.P. (2002) Immunobiology of tumor necrosis factor receptor superfamily. *Immunol. Res.* **26**, 323–336
2. Bodmer, J.L., Schneider, P. & Tschopp, J. (2002) The molecular architecture of the TNF superfamily. *Trends Biochem. Sci.* **27**, 19–26
3. Fesik, S.W. (2000) Insights into programmed cell death through structural biology. *Cell* **103**, 273–282
4. Chan, F.K., Chun, H.J., Zheng, L., Siegel, R.M., Bui, K.L. & Lenardo, M.J. (2000) A domain in TNF receptors that mediates ligand-independent receptor assembly and signaling. *Science* **288**, 2351–2354
5. Sartorius, U., Schmitz, I. & Krammer, P.H. (2001) Molecular mechanisms of death-receptor-mediated apoptosis. *Chembiochem.* **2**, 20–29
6. Grell, M., Zimmermann, G., Gottfried, E., Chen, C.M., Grunwald, U., Huang, D.C.S., Lee, Y.H.W., Durkop, H., Engelmann, H., Scheurich, P. et al. (1999) Induction of cell death by tumour necrosis factor (TNF) receptor 2, CD40 and CD30: a role for TNF-R1 activation by endogenous membrane-anchored TNF. *EMBO J.* **18**, 3034–3043
7. Trauth, B.C., Klas, C., Peters, A.M., Matzku, S., Moller, P., Falk, W., Debatin, K.M. & Krammer, P.H. (1989) Monoclonal antibody-mediated tumor regression by induction of apoptosis. *Science* **245**, 301–305
8. Yonehara, S., Ishii, A. & Yonehara, M. (1989) A cell-killing monoclonal antibody (anti-Fas) to a cell surface antigen co-downregulated with the receptor of tumor necrosis factor. *J. Exp. Med.* **169**, 1747–1756
9. Wajant, H., Pfizenmaier, K. & Scheurich P. (2003) Non-apoptotic Fas signaling. *Cytokine Growth Factor Rev.* **14**, 53–66
10. Wajant, H., Pfizenmaier, K. & Scheurich, P. (2002) TNF-related apoptosis inducing ligand (TRAIL) and its receptors in tumor surveillance and cancer therapy. *Apoptosis* **7**, 449–459
11. Martinon, F., Hofmann, K. & Tschopp, J. (2001) The pyrin domain: a possible member of the death domain-fold family implicated in apoptosis and inflammation. *Curr. Biol.* **11**, R118–R120
12. Donepudi, M., Sweeney, A.M., Briand, C. & Grutter, M.G. (2003) Insights into the regulatory mechanism for caspase-8 activation. *Mol. Cell.* **11**, 543–549
13. Boatright, K.M., Renatus, M., Scott, F.L., Sperandio, S., Shin, H., Pedersen, I.M., Ricci, J.E., Edris, W.A., Sutherlin, D.P., Green, D.R. & Salvesen, G.S. (2003) A unified model for apical caspase activation. *Mol. Cell* **11**, 529–554

14. Shi, Y. (2002) Apoptosome: the cellular engine for the activation of caspase-9. *Structure* **10**, 285–288

15. Holler, N., Zaru, R., Micheau, O., Thome, M., Attinger, A., Valitutti, S., Bodmer, J.L., Schneider, P., Seed, B. & Tschopp, J. (2000) Fas triggers an alternative, caspase-8-independent cell death pathway using the kinase RIP as effector molecule. *Nat. Immunol.* **1**, 489–495

16. Thome, M. & Tschopp, J. (2001) Regulation of lymphocyte proliferation and death by FLIP. *Nat. Rev. Immunol.* **1**, 50–58

17. Micheau, O., Thome, M., Schneider, P., Holler, N., Tschopp, J., Nicholson, D.W., Briand, C. & Grutter, M.G. (2002) The long form of FLIP is an activator of caspase-8 at the Fas death-inducing signaling complex. *J. Biol. Chem.* **277**, 45162–45171

18. Siegmund, D., Mauri, D., Peters, N., Juo, P., Thome, M., Reichwein, M., Blenis, J., Scheurich, P., Tschopp, J. & Wajant, H. (2001) Fas-associated death domain protein (FADD) and caspase-8 mediate up-regulation of c-Fos by Fas ligand and tumor necrosis factor-related apoptosis-inducing ligand (TRAIL) via a FLICE inhibitory protein (FLIP)-regulated pathway. *J. Biol. Chem.* **276**, 32585–32590

19. Nagata, S. & Suda, T. (1995) Fas and Fas ligand: lpr and gld mutations. *Immunol. Today* **16**, 39–43

20. Cretney, E., Takeda, K., Yagita, H., Glaccum, M., Peschon, J.J. & Smyth, M.J. (2002) Increased susceptibility to tumor initiation and metastasis in TNF-related apoptosis-inducing ligand-deficient mice. *J. Immunol.* **168**, 1356–1361

21. Sedger, L.M., Glaccum, M.B., Schuh, J.C., Kanaly, S.T., Williamson, E., Kayagaki, N., Yun, T., Smolak, P., Le, T., Goodwin, R. & Gliniak, B. (2002) Characterization of the in vivo function of TNF-alpha-related apoptosis-inducing ligand, TRAIL/Apo2L, using TRAIL/Apo2L gene-deficient mice. *Eur. J. Immunol.* **32**, 2246–2254

22. Lamhamedi-Cherradi, S.E., Zheng, S.J., Maguschak, K.A., Peschon, J. & Chen, Y.H. (2003) Defective thymocyte apoptosis and accelerated autoimmune diseases in TRAIL-/- mice. *Nat. Immunol.* **4**, 255–260

23. Wajant, H., Pfizenmaier, K. & Scheurich P. (2003) Tumor necrosis factor signaling. *Cell Death Differ.* **10**, 45–65

24. Karin, M. & Lin, A. (2002) NF-kappaB at the crossroads of life and death. *Nat. Immunol.* **3**, 221–227

25. Wajant, H. & Scheurich, P. (2001) Tumor necrosis factor receptor-associated factor (TRAF) 2 and its role in TNF signaling. *Int. J. Biochem. Cell Biol.* **33**, 19–32

26. Wang, E.C., Thern, A., Denzel, A., Kitson, J., Farrow, S.N. & Owen, M.J. (2001) DR3 regulates negative selection during thymocyte development. *Mol. Cell Biol.* **21**, 3451–3461

27. Zhao, H., Yan, M., Wang, H., Erickson, S., Grewal, I.S. & Dixit, V.M. (2001) Impaired c-Jun amino terminal kinase activity and T cell differentiation in death receptor 6-deficient mice. *J. Exp. Med.* **194**, 1441–1448

28. Liu, J., Na, S., Glasebrook, A., Fox, N., Solenberg, P.J., Zhang, Q., Song, H.Y. & Yang, D.D. (2001) Enhanced CD4+ T cell proliferation and Th2 cytokine production in DR6-deficient mice. *Immunity* **15**, 23–34

29. Roux, P.P. & Barker, P.A. (2002) Neurotrophin signaling through the p75 neurotrophin receptor. *Prog. Neurobiol.* **67**, 203–233

30. Thesleff, I. & Mikkola, M.L. (2002) Death receptor signaling giving life to ectodermal organs. *Sci. STKE* **131**, PE22

6

Guardians of cell death: the Bcl-2 family proteins

Peter T. Daniel*[1], Klaus Schulze-Osthoff†,
Claus Belka‡ and Dilek Güner§

*Molecular Hematology and Oncology, Charité - Campus Berlin-Buch,
Humboldt University, Lindenberger Weg 80, 13125 Berlin-Buch,
Germany, †Institute of Molecular Medicine, Heinrich Heine University,
Düsseldorf, Germany, ‡Department of Radiation Oncology, Eberhard
Karls University, Tübingen, Germany, and §Department of Radiation
Oncology, University Medical Center Charité, Humboldt University,
Berlin, Germany

Abstract

Apoptosis is mediated through at least three major pathways that are regulated
by (i) the death receptors, (ii) the mitochondria and (iii) the ER (endoplasmic
reticulum). In most cells, these pathways are controlled by the Bcl-2 family of
proteins that can be divided into anti-apoptotic and pro-apoptotic members.
Although the overall amino acid sequence homology between the family
members is relatively low, they contain highly conserved domains, referred to
as BH (Bcl-2 homology) domains (BH1–4), that are essential for homo- and
hetero-complex formation, as well as for their cell-death-inducing capacity.
Structural and functional analyses revealed that the pro-apoptotic homologues
can be subdivided into the Bax subfamily and the growing BH3-only
subfamily. Recent data indicate that BH3-only proteins act as mediators that
link various upstream signals, including death receptors and DNA damage

[1]To whom correspondence should be addressed (e-mail pdaniel@mdc-berlin.de).

signalling, to the mitochondrial and the ER pathway. This review discusses recent structural and functional insights into how these subfamilies promote or inhibit cell-death signals, and how these properties may be utilized for development of apoptosis-promoting small molecules, e.g. in cancer therapy.

Introduction

Members of the *bcl-2* (the B-cell lymphoma gene 2) gene family play a key role as sensors and gatekeepers that control the execution machinery of apoptosis. Like the cysteinyl aspartases (caspases) that mediate the initiation and execution of apoptosis, the evolutionary origins of the constituents of the apoptosome, a key regulatory protein complex that mediates cell death through the mitochondrial pathway, can be traced back to *Caenorhabditis elegans* and *Drosophila*.

The importance of Bcl-2 and related molcules becomes obvious when the phenotypes of mutant animals are analysed. Targeted gene-knockout of the *bcl-2* gene in mice results in the occurrence of grey hair, polycystic kidney disease and lymphocytopenia. In contrast, homozygous disruption of the *bcl-2*-homologous *bcl-x* gene is lethal during embryonic development and coincides with massive apoptosis in postmitotic immature neurons of brain and spinal cord and in the haematopoietic system. Analyses of mice carrying only one disrupted *bcl-x* allele showed that lymphocyte maturation was disturbed. The life-span of immature lymphocytes, but not mature lymphocytes, was shortened. Thus, Bcl-x functions to support the viability of immature cells during the development of the nervous and haematopoietic systems. Knockout of the pro-apoptotic *bcl-2*-homologous *bax* gene (see below) in mice results in a milder phenotype, including hyperplasia of non-neuronal lineages, such as lymphocytes and ovarian granulosa cells, and testicular degeneration. In the nervous system, Bax is required for neuronal death after deprivation of neurotrophic factors. Interestingly, loss of Bax [1] or high expression of Bcl-2 [2] facilitates tumorigenesis. Thus deregulation of Bcl-2 homologues and consecutive resistance to apoptosis appears to be a critical event in oncogenic transformation.

Recent studies have provided novel and quite unexpected insights into the mechanisms by which Bcl-2 family proteins might issue life permits or death sentences in cells. Nevertheless, we are still on the way to fully understanding their modes of action. This review compiles recent developments regarding the structural and functional regulation of these proteins and how we may exploit our knowledge of this family of proteins to fight diseases such as cancer or auto-immunity, where loss of apoptosis control is critically involved in pathogenesis.

Regulation of apoptosis by Bcl-2 family proteins

Although *bcl-2*, the first gene implicated in the control of apoptosis, was identified in 1984 [3], the way in which Bcl-2 and its homologues regulate cell death is still far from being elucidated. Insights into the regulatory functions of

Bcl-2 and its homologues were gained when it was recognized that mitochondria play a critical role in apoptosis, whether the extrinsic, death-receptor-mediated, or the intrinsic apoptosis pathway is triggered, e.g. upon nuclear stress (Figure 1). Upon activation during apoptosis, they release proteins such as cytochrome c, Smac (second mitochondrial activator of

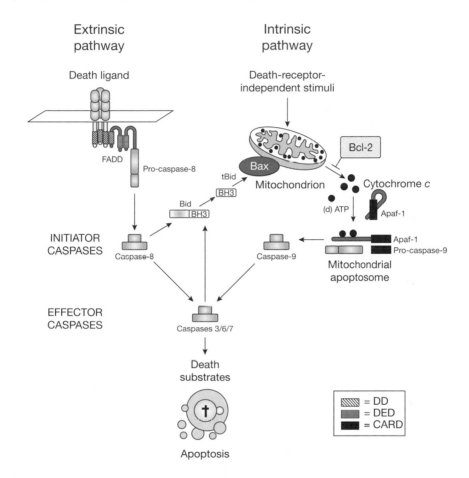

Figure 1. Bcl-2 proteins regulate the intrinsic and extrinsic apoptosis pathways

The extrinsic apoptosis pathway (left) is induced by death receptor ligands, such as the CD95/Fas ligand (CD95L). A death-inducing signalling complex consisting of CD95L, the CD95 receptor, the adapter protein FADD and pro-caspase-8 is formed at the cell membrane, leading to activation of the initiator caspase-8. Caspase-8 in turn cleaves and activates the effector caspase-3. In contrast, the intrinsic death receptor-independent pathway (right) is triggered by the majority of apoptotic stimuli, including cytotoxic drugs, ceramide and irradiation, and leads to the loss of the mitochondrial membrane potential and release of cytochrome c and other pro-apoptotic factors (see text) into the cytosol. Bcl-2 inhibits, whereas pro-apoptotic Bcl-2 homologues, such as Bax or Bak, promote release of these pro-apoptotic factors. Cytochrome c, together with (d)ATP, Apaf-1 and pro-caspase-9, then forms the mitochondrial apoptosome leading to activation of the initiator caspase-9, followed by activation of the effector caspases 3, 6 and 7. This pathway can be amplified by caspase-8- or -3-mediated cleavage of Bid to tBid that then activates Bax. Apart from mediating this auto-amplification loop, Bid connects the extrinsic to the intrinsic apoptosis pathway. DD, death domain; DED, death-effector domain; CARD, caspase-recruitment domain.

caspases), AIF (apoptosis-inducing factor), the serine protease Omi/Htr2 and others, such as endonuclease G, into the cytosol [4]. This, and changes in mitochondrial membrane permeability leading to efflux of H^+, Ca^{2+} and other ion fluxes that cause acidification of the cytosol and loss of the mitochondrial membrane potential, are early events during apoptosis. Cytochrome c forms a complex with the adaptor protein Apaf-1 (apoptotic protease-activating factor 1), dATP or ATP, and pro-caspase-9 in the cytosol. Upon formation of this 'apoptosome', the caspase-9 zymogen is processed and triggers the cleavage and activation of effector caspases, ultimately leading to cell death. These events are counteracted by Bcl-2 and its anti-apoptotic homologues, such as Bcl-x$_L$ [5]. Both Bcl-2 and Bcl-x$_L$ inhibit the release of cytochrome c and other mitochondrial events during apoptosis.

Interestingly, some of these Bcl-2-homologous proteins do not inhibit apoptosis, but on the contrary, promote cell death. The Bcl-2-associated x protein Bax was the first Bcl-2 homologue to be identified that had apoptosis-promoting properties. Bax is a direct activator of mitochondria that triggers the release of cytochrome c [6] and other mitochondrial events, such as the breakdown of the mitochondrial membrane potential and the release of pro-apoptotic factors.

Structure, function and protein interactions

The search for *bcl-2*-interacting genes by means of yeast two-hybrid screens and for *bcl-2*-homologous genes by PCR strategies based on the use of degenerate primers rapidly yielded a large number of *bcl-2*-homologous genes. A list and classification based on structure and function of the growing family of anti- and pro-apoptotic *bcl-2* homologues is given in Figure 2. Such structural and functional analyses showed that the growing family of *bcl-2*-homologous genes can be divided into at least three distinct subfamilies: the Bcl-2 and the Bax subfamily, and the BH3-only proteins (where BH means Bcl-2 homology). DNA and protein sequence alignments identified five homologous conserved regions in Bcl-2 family members that are critical for function and protein–protein interaction: four Bcl-2-homology regions, BH1 (amino acids 136–155), BH2 (187–202), BH3 (93–107) and BH4 (10–30), plus a transmembrane region. All cellular anti-apoptotic Bcl-2 homologues carry a BH1, a BH2 and a BH4 domain. In contrast, the *C. elegans* homologues and related proteins found in pathogenic viruses show high sequence homology only in the BH1 and BH2 domains, whereas functions of the BH4 (and BH3) domains appear to be carried out by non-homologous regions of the proteins. Homology of human Bcl-2 homologues with *C. elegans* proteins demonstrates the evolutionary conservation and helped to identify functionally relevant domains. Moreover, the presence of anti-apoptotic *bcl-2*-homologous genes in pathogenic viruses delineates the functional relevance of apoptosis in the control of virus infection and limitation of viral replication.

Notably, Bcl-2 homologues can interact with each other. These interactions occur through the BH1, the BH2 and, most importantly, the BH3 domain. Deletion of the BH1, BH2 or BH4 domains of Bcl-2 impairs its ability to suppress cell death in mammalian cells and prevents homodimerization of these mutant proteins, although they can still bind to the wild-type Bcl-2 protein. In contrast, all pro-apoptotic family members contain a highly conserved BH3 domain.

Further analyses revealed that the apoptosis-promoting Bcl-2 family members can be organized into two subfamilies: (i) the Bax homologues, including Bax, Bak and Bok/Mtd, and (ii) a subfamily that includes Bad, Bid, Bim, Bmf, Hrk, Nbk, Bnip3, Noxa and Puma [7]. These homologues carry only one of the signature domains of Bcl-2, the BH3 domain, and are therefore designated BH3-only proteins. Interestingly, mutation of the BH3 domain destroys the apoptotic function of Bcl-2 homologues. In addition, cell-permeant peptides derived from Bax or other pro-apoptotic Bcl-2 homologues antagonize Bcl-2- and Bcl-x_L-mediated protection from apoptosis. Surprisingly, exchanging the

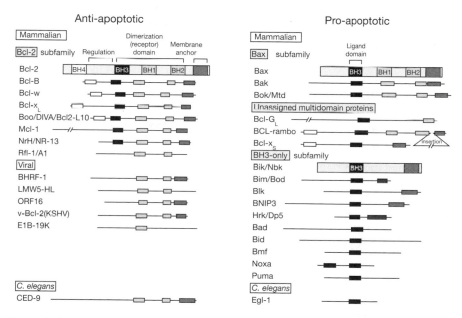

Figure 2. Structural and functional classification of Bcl-2 family proteins
Structural analyses identified five homologous regions in Bcl-2 family members. These are the BH domains BH1, BH2 (light-blue boxes), BH3 (dark blue) and BH4 (white), and the transmembrane domain (mid-blue boxes). Anti-apoptotic viral and *C. elegans* Bcl-2 homologues do not carry a BH3 and BH4 domain. The BH3 domain in the pro-apoptotic members binds to the hydrophobic pocket formed by the BH1–BH3 domains of the survival factors. The C-terminal hydrophobic domain mediates anchoring of some of the proteins in intracellular membranes. Structure–function analyses identified three distinct subfamilies: the anti-apoptotic Bcl-2 homologues, the pro-apoptotic multidomain Bax homologues and the subfamily of BH3-only proteins. The latter act as indirect mediators of cell death through interaction with the Bax-homologous multidomain pro-apoptotic proteins. A number of additional pro-apoptotic multidomain proteins show distinct structural and functional characteristics, such as Bcl-G_L, or carry a BH4 domain, such as Bcl-x_S, and were therefore not yet assigned to a distinct pro-apoptotic subfamily.

BH3 domains between Bcl-2 and Bax did not impair the propensity of Bax to induce apoptosis. In fact, Bcl-2 may be converted to a pro-apoptotic molecule with Bax-like functions during apoptosis. This occurs through proteolysis of Bcl-2 by caspase-3 that eliminates the N-terminal part containing the BH4 and the loop domains. Such cleavage-resistant mutants of Bcl-2 mediated increased protection from apoptosis. Thus cleavage of Bcl-2 by caspases may ensure the inevitability of cell death.

In addition, these data indicate that other parts of the Bcl-2 molecule and its anti-apoptotic homologues apart from the BH1–BH3 domains are involved in mediating their death-inhibitory properties. Analyses of three-dimensional solution structures and functional analyses of deletion mutants revealed that this death-inhibitory function must be located in the N-terminal regions of the protein containing the BH4 domain. The conserved BH4 domain is found in all anti-apoptotic Bcl-2 homologues.

Previous reports have suggested at least two distinct mechanisms for inhibition of apoptosis: Bcl-2 and Bcl-x_L might inhibit either the formation of the cytochrome c/Apaf-1/caspase-9 apoptosome complex by preventing cytochrome c release from mitochondria or they might interfere with the function of this apoptosome through a direct interaction of Bcl-2 or Bcl-x_L with Apaf-1.

The three-dimensional solution structure of the full-length Bax protein has been solved by NMR [8]. The Bax structure shows a high similarity to the overall conformation of the four other Bcl-2 family proteins for which structural information is available, the anti-apoptotic proteins Bcl-w, Bcl-x_L and Kaposi's sarcoma virus v-Bcl-2, and the pro-apoptotic BH3-only protein Bid. The proteins contain central hydrophobic helices (α5 and α6) surrounded by amphipathic helices with some (minor) structural differences in the v-Bcl-2. The BH1, BH2 and BH3 domains are located in close proximity on the surface of the protein and form a hydrophobic groove that also contains the BH3 domain and appears to represent the binding site for other Bcl-2 family members [9]. Structural analyses indicate that activation of Bax homologues results in conformational changes that expose the hydrophobic transmembrane domain and thereby trigger insertion into membranes. This coincides with exposure of the BH3-domain-containing groove that would then allow homo- and hetero-philic interactions, i.e. oligomerization and channel formation in the case of Bax and Bak.

Some reports suggested that the BH4 domain is critical for binding of Bcl-x_L to the apoptosome. It was postulated that Bcl-x_L (and homologues) would prevent apoptosis by interfering with the activation of pro-caspase-9 in the complex formed with Apaf-1 and cytochrome c. The physical interaction of Bcl-x_L is, however, rather weak and no physical interaction was found in co-immunoprecipitation studies with anti-apoptotic Bcl-2 homologues (Bcl-2, Bcl-x_L, Bcl-w, A1, Mcl-1 or Boo), viral Bcl-2 mimics (adenovirus E1B 19K and Epstein–Barr virus BHRF-1) or with two pro-apoptotic relatives (Bax and

Bim) [10]. Furthermore, recombinant Bcl-x$_L$ inhibited mitochondrial activation and cytochrome release when added to isolated mitochondria in a cell-free system, but failed to inhibit caspase activation induced by exogenous cytochrome c and dATP in cytosolic extracts, i.e. occuring downstream of the mitochondria [11].

Subcellular localization, redistribution and channel formation

Novel insights into the mode of action of Bcl-2 family proteins also came from studies investigating their subcellular-distribution patterns. Bcl-2 was shown early on to locate to the outer mitochondrial membrane. Bcl-2 is also found at the nuclear and the ER (endoplasmic reticulum) membranes facing the cytosol. In contrast, Bcl-x$_L$ was shown recently to localize preferentially to the mitochondria. The pro-apoptotic Bcl-2 homologues Bax and Bak may localize both to the ER and the mitochondria.

The subcellular distribution of splice variants of these proteins may differ. Bcl-2 can occur as Bcl-2α or Bcl-2β, two alternatively spliced forms that differ solely in their C-termini. The finding that Bcl-2α is active and membrane-bound, but that Bcl-2β is inactive and cytosolic, indicates that the C-terminus and the membrane localization contributes to the survival activity of Bcl-2α [12]. Similarly, the short splice variant of Bcl-x, Bcl-x$_S$, lacks a transmembrane domain and is found in the cytosol whereas Bcl-x$_L$ is found in both the cytosol and the outer mitochondrial membrane in healthy, non-apoptotic cells. During apoptosis, Bcl-x$_L$ redistributes quantitatively to the mitochondria. Thus the membrane localization seems to be critical for the function of anti-apoptotic Bcl-2 homologues. Likewise, Bax is found as a monomer in the cytosol of most normal healthy tissues, which indicates that the protein is kept in an inactive form in the cells [13]. During induction of the mitochondrial pathway, Bax and Bak become activated and undergo a conformational switch in the N-terminus and the C-terminus that results in redistribution of Bax to the mitochondria [14]. Activation of Bax results in insertion of Bax into the outer mitochondrial membrane. Bax translocation to the mitochondria coincides with oligomerization of Bax to multimeric complexes. Likewise, recombinant Bax oligomerizes and forms channels in artificial liposomal membranes [15]. Such Bax oligomers possess channel-forming properties that are associated with ion fluxes and consecutive loss of the mitochondrial membrane potential. In contrast, Bak resides constitutively in the mitochondrial membrane and the ER (see below), where it oligomerizes upon activation [16].

Channel formation by Bax or Bak oligomers has been proposed to be responsible for the release of cytochrome c from the mitochondria. In fact, image analysis of the interaction of the Bax with membrane bilayers shows the presence of toroidal-shaped pores by means of atomic force microscopy. These pores are large enough to allow for passage of proteins such as cytochrome c from the intermembrane space of mitochondria.

Nevertheless, the release of cytochrome c release through Bax or Bak channels is discussed controversially. Activation of other channels, such as the permeability transition pore, has been implicated in the release of cytochrome c and ion fluxes induced by Bax and Bak. This channel is formed from a complex of the VDAC (voltage-dependent anion channel), the ANT (adenine nucleotide translocase) and cyclophilin-D at contact sites between the mitochondrial outer and inner membranes. There is some evidence that this complex can recruit a number of other proteins, including Bax [17]. Thus Bax and other Bcl-2 homologues have been implicated as regulators of the VDAC/ANT permeability pore. There is, however, also evidence that Bax oligomers do, at best, bind only weakly to VDAC channels.

The situation for other pro-apoptotic factors released from mitochondria is far less clear. Recent data suggest that Smac is also released through a Bax/Bak-dependent mechanism while others observed temporary dissociation of cytochrome c and Smac release, indicating that distinct mechanisms might be involved. Nevertheless, pro-apoptotic factors may also be released during late mitochondrial activation, when ion fluxes through the permeability transition pore formed by VDAC proteins in the outer and ANT in the inner mitochondrial membrane (and so-far-undefined 'megapores') have led to swelling and fragmentation of the mitochondria, a mechanism that can be prevented by Bcl-x_L. Thus AIF, and possibly other factors appear, to be released through mechanisms that operate downstream of cytochrome c and are independent from Bax [18].

Recently, an interaction between Bax and Drp1 (dynamin-related protein 1) was described that suggests a completely different view of how Bcl-2 and its homologues regulate mitochondrial apoptosis [19]: clustering of mitochondria around the nucleus and mitochondrial fragmentation is a typical feature of cells undergoing apoptosis. Interestingly, Bax translocates to discrete foci on mitochondria during the initial stages of apoptosis, which subsequently become mitochondrial scission sites. Surprisingly, the dynamin-related GTPase Drp1 and the mitofusin GTPase Mfn2, but not other proteins implicated in the regulation of mitochondrial morphology and physiological fission and fusion events, co-localize with Bax in these foci. Mfn2 is an outer mitochondrial membrane protein with N-terminal and C-terminal domains exposed towards the cytosol. Interestingly, Mfn2 over-expression can trigger perinuclear clustering of mitochondria, a phenomenon that is frequently observed during apoptosis. Co-expression of Mfn2 with a dominant-interfering mutant Drp1 proposed to block mitochondrial fission results in long mitochondrial filaments and networks. Moreover, a dominant-negative mutant of Drp1 inhibited apoptotic scission of mitochondria, but did not inhibit Bax translocation or coalescence into foci. This indicates that Bax participates in apoptotic fragmentation of mitochondria through interaction with

Drp1/Mfn2. The question whether this mechanism is the real cause for triggering the mitochondrial apoptosis signalling events remains, however, to be clarified. A compilation of the various models and levels of regulation is provided in Figure 3.

Finally, Bcl-2 homologues might act in an organelle-specific manner [20]. Whereas Bcl-2 is distributed in a rather promiscuous manner to cellular organelle membranes facing the cytosol, activated Bcl-x_L preferentially localizes to mitochondria. Likewise, the BH3-only protein Nbk (natural born killer; or Bik, Bcl-2-interacting killer) distributes to the ER and not to the mitochondria, indicating that Nbk triggers Bax activation in the mitochondria through an indirect mechanism. This view is supported by the observation that Nbk induces apoptosis in a Bax-dependent manner, but fails to directly bind to Bax [21]. Moreover, Bak localizes to the ER where it regulates Ca^{2+} efflux that might trigger secondary activation of mitochondria following ER stress responses. Likewise, Bcl-2 expression in the ER inhibits Ca^{2+} release from the ER calcium pool and inhibits apoptosis upon various stimuli [22]. This indicates a broader role for the ER in apoptosis control than previously imagined

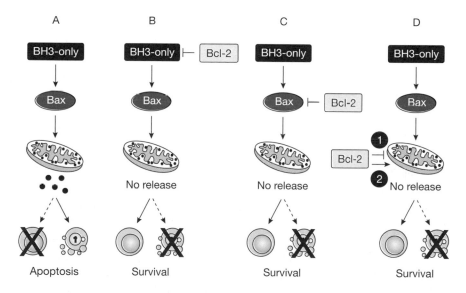

Figure 3. Models for the pro-survival activity of Bcl-2
(**A**) Recent data established that BH3-only proteins initiate the mitochondrial pathway of apoptosis through the triggering of pro-apoptotic Bax or its homologues. Activation of Bax by BH3-only proteins triggers a conformational switch, insertion into cellular membranes and Bax channel formation. This results in the release of pro-apoptotic factors from the mitochondria and initiation of caspase-dependent or -independent pathways that execute apoptosis. The following models of how Bcl-2 inhibits the mitochondrial pathway are matter of current debate: (**B**) sequestration of BH3-only proteins by Bcl-2 that serves here as a sink; (**C**) inhibition of Bax activation; (**D**) interference with mitochondrial events, such as opening of VDAC or other channels and release of pro-apoptotic factors. Another possibility is the interference with mitochondrial fission (see text). It remains to be established whether these effects are mediated through (i) inhibition of pro-apoptotic events or (ii) direct stabilization. The interference with downstream factors such as Apaf-1 or the caspases has been formally excluded.

that might equal that of the mitochondria [20]. Last, but not least, given their distinct roles in the ER and the mitochondria, there is accumulating evidence that Bax and Bak might not be fully redundant in regulating apoptosis. In fact, the BH3-only protein Nbk triggers apoptosis in an entirely Bax-dependent, but Bak-independent, manner [21]. This adds a higher degree of specificity and complexity to the previously suggested interchangeability of Bcl-2 homologues in the regulation of apoptosis.

BH3-only proteins as sensors that mediate activation of Bax and Bak

Recently, it was demonstrated that the pro-apoptotic activities of certain BH3-only proteins essentially depends on the presence of either Bax or Bak (reviewed in [7]). This finding suggests that some BH3-only proteins can promote apoptosis by two mutually non-exclusive mechanisms: by directly inactivating pro-survival Bcl-2-proteins or through direct or indirect modification of Bax-like molecules.

BH3-only proteins may function as death sensors and their pro-apoptotic activity is subject to stringent control (Figure 4). For example, Noxa, Puma, Nbk and Hrk are transcriptionally induced by p53 upon stress conditions. Many other BH3-only proteins are present in healthy cells in a dormant form and are activated upon post-translational modifications, such as phosphorylation of Bad or Nbk. Bim and Bmf represent two BH3-only proteins that are regulated by sequestration to cytoskeletal structures. During apoptosis mediated upon ligation of death receptors, the BH3-only protein Bid is proteolytically cleaved by caspase-8 to a pro-apoptotic form (tBid) from an inactive form. tBid appears to interact physically with Bax to mediate a conformational change in the Bax N-terminus [23]. This event triggers insertion of cytosolic Bax into the outer mitochondrial membrane [14]. Thus Bid links death-receptor signalling to the mitochondrial pathway. Interestingly, Bid cleavage can occur in a death receptor-independent manner downstream of the mitochondria through a caspase-3- and -8-dependent mechanism [24]. Thus caspase-3, caspase-8 and Bid cleavage mediate a positive-feedback loop that is required for full cytochrome c release during anti-cancer drug-induced apoptosis [25]. Other reports found that Bid can also be cleaved by non-caspase proteases, especially granzyme B and lysosomal proteases, e.g. the cathepsins. Apart from mediating the above-mentioned amplification loop, Bid cleavage by lysosomal proteases was implicated in cross-talk between lysosomes and mitochondria during apoptosis.

Unlike several other members of the BH3-only family, Nbk/Bik triggers apoptosis in an entirely Bax-dependent manner [21]. Ectopic expression of Nbk impaired the tumorigenicity in a SCID (severe combined inmmunodeficient) mouse xenotransplant model and promoted drug-induced apoptosis, indicating that Nbk acts in the DNA-damage-response pathway [26]. It has been proposed that the activity of the BH3-only protein Nbk is controlled by phosphorylation. Opposite to the case of Bad, phosphorylation increases the pro-apoptotic potency of Bik through a presently unknown mechanism that does not affect its affinity for anti-apoptotic Bcl-2 members.

Like other BH3-only proteins, including Nbk, the BH3-only protein Bad interacts via its BH3 domain with Bcl-x_L. This may indicate that anti-apoptotic Bcl-2 homologues serve as a 'sink' that sequesters BH3-only proteins either in the cytosol or at membranes. In fact, Bad is inactivated upon phosphorylation of Ser-155 within the BH3 domain through the anti-apoptotic Akt kinase [27]. This leads to sequestration of the phosphorylated Bad in the cytosol through binding to 14-3-3 proteins. The sequestration of Bad, in turn, was shown to enhance the anti-apoptotic properties of Bcl-x_L. This appears to occur through phosphorylation and binding of Bad to 14-3-3 proteins instead of binding of the unphosphorylated Bad to Bcl-x_L. Thus phosphorylation of Bad impairs its pro-apoptotic potency.

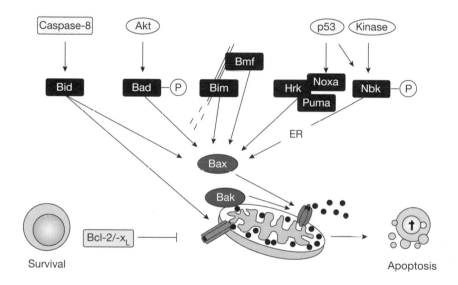

Figure 4. Function of BH3-only proteins as death sensors
BH3-only proteins link upstream signals from different cellular or functional compartments to the mitochondrial apoptosis pathway. Cleavage of Bid by caspases 8 or 3, e.g. during death-receptor-induced apoptosis, generates tBid that binds to Bax. This triggers a conformational switch in Bax that exposes the Bax C-terminus, induces redistribution of Bax to the mitochondria and insertion into the outer mitochondrial membrane. In contrast, Bak is localized constitutively in the outer mitochondrial membrane. Upon activation, both Bax and Bak oligomerize and form channels that release cytochrome c. Bim is released from the motor-dynein complex, whereas Bmf is released from actin–myosin filaments during apoptosis. Puma, Noxa, Hrk and Nbk (Bik) are induced by p53 and mediate cell death originating from the nucleus, e.g. upon DNA damage. Phosphorylation of Nbk by an as-yet-undefined kinase enhances Nbk function. Nbk activates Bax indirectly, through an ER-dependent pathway. In contrast, phosphorylation of Bad through Akt/protein kinase B inactivates Bad. The latter disrupts the binding of Bad to Bcl-x_L and results in Bax- and Bak-dependent apoptosis. Apart from channel formation through Bax and Bak, cytochrome c can be released through activation of VDAC channels or oligomerization of Bid in the outer mitochondrial membrane (see text).

Finally, Bcl-2 is subject to post-translational modification during apoptosis, mainly upon exposure of cells to microtubule toxins such as vinca alkaloids and taxanes that trigger unco-ordinated microtubule polymerization. Hyperphosphorylation of Bcl-2 mediated by Paclitaxel and other microtubule-active drugs is strictly dependent on targeting microtubules that in turn cause mitotic arrest. In addition to Ser-70, microtubule-active agents promote phosphorylation of Ser-87 and Thr-69, thereby inactivating Bcl-2. This multi-site phosphorylation of the Bcl-2 loop domain is followed by cell death. In contrast, in cytokine-dependent cell lines, cytokines mediate phosphorylation of Bcl-2 at Ser-70 that prevents apoptosis [28]. Thus functions of Bcl-2 family proteins are differentially regulated by serine/threonine phosphorylation and this seems to be of special importance at the level of the BH3-only proteins.

Bcl-2 family proteins as drug targets in cancer

Over-expression of Bcl-2, Bcl-x_L or other anti-apoptotic homologues or loss of Bax and Bak is a common feature of human tumours. Apart from being involved in tumorigenesis where deregulation of Bcl-2 homologues is required to counteract pro-apoptotic properties of oncogenes, expression and mutation analyses in human tumours revealed that over-expression of anti-apoptotic Bcl-2 family members or loss of Bax homologues is associated with the acquisition of clinical resistance to anti-cancer therapies [29,30]. Interestingly, clinical data show that the loss of pro-apoptotic homologues, rather than the high expression of anti-apoptotic homologues is associated with resistance to anti-cancer therapy [31,32].

The manipulation of apoptosis for therapeutic purposes is receiving considerable attention from the pharmaceutical industry. In fact, a variety of reports demonstrated that the enforced expression of pro-apoptotic Bcl-2 homologues facilitates apoptosis, decreases tumorigenicity, and overcomes resistance to chemotherapy and ionizing irradiation [26]. This suggested that Bcl-2 proteins are putative targets for development of novel anti-tumour agents (reviewed in [33]). Based on structural and functional data, antagonists of Bcl-2 were developed based on different strategies. Antisense oligonucleotides against Bcl-2 are currently evaluated in phase III multi-centre clinical trials as anti-cancer agents in malignant melanoma, follicular lymphoma and multiple myeloma. Results from phase I/II trials showed good activity and tumour regression in some patients.

Nevertheless, the Bcl-2 family consists of proteins with redundant functions. Thus targeting multiple Bcl-2 homologues at the same time appears to be more promising. This was achieved *in vitro* by the use of cell-permeant BH3-only domain-derived peptides that potently trigger apoptosis. Recently, a number of small molecules were developed as BH3 mimics that bind to the BH3-containing cleft in Bcl-2 or Bcl-x_L. These small-molecule inhibitors specifically block the BH3-domain-mediated heterodimerization between

Bcl-2 family members *in vitro* and *in vivo* and consequently induce apoptosis (reviewed in [33]). At least one of these Bcl-2-inhibitory compounds acts through the activation of Bax. Surprisingly, loss of the pro-apoptotic Bax or overexpression of Bcl-2 results in a selective resistance against some, but not all, cancer therapeutics, thereby indicating that some, but not all, cancer therapies depend on Bax or Bcl-2 to execute cell death [34]. This, together with the frequent inactivation of Bax and/or Bak in cancer, indicates that Bcl-2-targeting agents, despite their apparent potency, might not yet be the 'magic bullets' urgently required to overcome resistance to therapy in cancer and to improve cure rates. Further insights into the structural differences between various Bcl-2 homologues and their BH3 domains should help to add a higher degree of specificity to such 'molecular' therapies and to avoid undesired toxicities.

Conclusions

Significant progress has been made in our understanding of how Bcl-2 proteins regulate apoptosis. Apart from regulating the mitochondrial and ER pathways of apoptosis, members of the Bcl-2 family functionally interconnect apoptosis pathways to converge at the mitochondria. Thus it is now clear that the subfamily of BH3-only proteins acts as sensor to trigger mitochondrial activation and amplification of the apoptotic signalling as a consequence of various upstream signals originating at the membrane, the nucleus, the cytoskeleton and other cellular compartments, including the ER. In this vein, the elucidation of three-dimensional solution structures of pro- and anti-apoptotic Bcl-2 proteins was a significant step towards a therapeutic manipulation of apoptosis signalling and the development of small-molecule inhibitors of Bcl-2 that should help to improve treatment of diseases with disturbed apoptosis signalling, such as cancer or autoimmunity.

Summary

- *Structural and functional analyses allow a classification of Bcl-2 family proteins in anti-apoptotic Bcl-2 homologues, and the two pro-apoptotic subfamilies of Bax homologues and BH3-only proteins.*
- *Bax and Bak are direct activators of mitochondria and trigger release of cytochrome c, and other pro-apoptotic proteins.*
- *Bax and Bak oligomerize in the outer mitochondrial membrane and form channels sufficiently large to release cytochrome c.*
- *In addition, Bax and Bak are implicated in the regulation of permeability pores formed by the VDAC in the mitochondrial membrane.*
- *BH3-only proteins link apoptosis signalling originating from various upstream events and cellular compartments to the mitochondria.*
- *BH3-only proteins mediate apoptosis through direct or indirect activation of Bax and Bak.*

- *Small molecules that mimic the BH3 domain of pro-apoptotic Bax or Bak are developed as drugs to inhibit Bcl-2 and to induce apoptosis in diseases with disrupted apoptosis signalling, e.g. cancer and auto-immune diseases.*

We thank the members of our groups for helpful discussions and their scientific contributions. Given the increasing complexity and the number of publications on the Bcl-2 family, we would like to apologize to those whose work could not be cited in this chapter.

References

1. Yin, C., Knudson, C.M., Korsmeyer, S.J. & Van Dyke, T. (1997) Bax suppresses tumorigenesis and stimulates apoptosis *in vivo*. *Nature (London)* **385**, 637–640
2. Strasser, A., Harris, A.W., Bath, M.L. & Cory, S. (1990) Novel primitive lymphoid tumours induced in transgenic mice by cooperation between myc and bcl-2. *Nature (London)* **348**, 331–333
3. Tsujimoto, Y., Finger, L.R., Yunis, J., Nowell, P.C. & Croce, C.M. (1984) Cloning of the chromosome breakpoint of neoplastic B cells with the t(14;18) chromosome translocation. *Science* **226**, 1097–1099.
4. van Loo, G., Saelens, X., van Gurp, M., MacFarlane, M., Martin, S.J. & Vandenabeele, P. (2002) The role of mitochondrial factors in apoptosis: a Russian roulette with more than one bullet. *Cell Death Differ.* **9**, 1031–1042
5. Newmeyer, D.D., Farschon, D.M. & Reed, J.C. (1994) Cell-free apoptosis in *Xenopus* egg extracts: inhibition by Bcl-2 and requirement for an organelle fraction enriched in mitochondria. *Cell* **79**, 353–364
6. Jurgensmeier, J.M., Xie, Z., Deveraux, Q., Ellerby, L., Bredesen, D. & Reed, J.C. (1998) Bax directly induces release of cytochrome *c* from isolated mitochondria. *Proc. Natl. Acad. Sci. U.S.A.* **95**, 4997–5002
7. Puthalakath, H. & Strasser, A. (2002) Keeping killers on a tight leash: transcriptional and post-translational control of the pro-apoptotic activity of BH3-only proteins. *Cell Death Differ.* **9**, 505–512.
8. Suzuki, M., Youle, R.J. & Tjandra, N. (2000) Structure of Bax: coregulation of dimer formation and intracellular localization. *Cell* **103**, 645–654
9. Muchmore, S.W., Sattler, M., Liang, H., Meadows, R.P., Harlan, J.E., Yoon, H.S., Nettesheim, D., Chang, B.S., Thompson, C.B. & Wong, S.L. (1996) X-ray and NMR structure of human Bcl-xL, an inhibitor of programmed cell death. *Nature (London)* **381**, 335–341
10. Moriishi, K., Huang, D.C., Cory, S. & Adams, J.M. (1999) Bcl-2 family members do not inhibit apoptosis by binding the caspase activator Apaf-1. *Proc. Natl. Acad. Sci. U.S.A.* **96**, 9683–9688
11. Newmeyer, D.D., Bossy-Wetzel, E., Kluck, R.M., Wolf, B.B., Beere, H.M. & Green, D.R. (2000) Bcl-xL does not inhibit the function of Apaf-1. *Cell Death Differ.* **7**, 402–407
12. Borner, C., Olivier, R., Martinou, I., Mattmann, C., Tschopp, J. & Martinou, J.C. (1994) Dissection of functional domains in Bcl-2 α by site-directed mutagenesis. *Biochem. Cell Biol.* **72**, 463–469
13. Hsu, Y.T. & Youle, R.J. (1998) Bax in murine thymus is a soluble monomeric protein that displays differential detergent-induced conformations. *J. Biol. Chem.* **273**, 10777–10783
14. Eskes, R., Desagher, S., Antonsson, B. & Martinou, J.C. (2000) Bid induces the oligomerization and insertion of Bax into the outer mitochondrial membrane. *Mol. Cell. Biol.* **20**, 929–935
15. Antonsson, B., Montessuit, S., Lauper, S., Eskes, R. & Martinou, J.C. (2000) Bax oligomerization is required for channel-forming activity in liposomes and to trigger cytochrome *c* release from mitochondria. *Biochem. J.* **345**, 271–278

16. Wei, M.C., Lindsten, T., Mootha, V.K., Weiler, S., Gross, A., Ashiya, M., Thompson, C.B. & Korsmeyer, S.J. (2000) tBID, a membrane-targeted death ligand, oligomerizes BAK to release cytochrome c. *Genes Dev.* **14**, 2060–2071

17. Crompton, M. (1999) The mitochondrial permeability transition pore and its role in cell death. *Biochem. J.* **341**, 233–249

18. Arnoult, D., Parone, P., Martinou, J.C., Antonsson, B., Estaquier, J. & Ameisen, J.C. (2002) Mitochondrial release of apoptosis-inducing factor occurs downstream of cytochrome c release in response to several proapoptotic stimuli. *J. Cell Biol.* **159**, 923–929

19. Frank, S., Gaume, B., Bergmann-Leitner, E.S., Leitner, W.W., Robert, E.G., Catez, F., Smith, C.L. & Youle, R.J. (2001) The role of dynamin-related protein 1, a mediator of mitochondrial fission, in apoptosis. *Dev. Cell* **1**, 515–525

20. Rudner, J., Jendrossek, V. & Belka, C. (2002) New insights in the role of Bcl-2: Bcl-2 and the endoplasmic reticulum. *Apoptosis* **7**, 441–447

21. Gillissen, B., Essmann, F., Graupner, V., Stärck, L., Radetzki, S., Dörken, B., Schulze-Osthoff, K. & Daniel, P.T. (2003) Induction of cell death by the BH3-only Bcl-2 homologue Bik/Nbk is mediated by a Bax-dependent mitochondrial pathway. *EMBO J.* **22**, 3580–3590

22. McConkey, D.J. & Nutt, L.K. (2001) Calcium flux measurements in apoptosis. *Methods Cell Biol.* **66**, 229–246

23. Desagher, S., Osen-Sand, A., Nichols, A., Eskes, R., Montessuit, S., Lauper, S., Maundrell, K., Antonsson, B. & Martinou, J. (1999) Bid-induced conformational change of bax is responsible for mitochondrial cytochrome c release during apoptosis. *J. Cell Biol.* **144**, 891–901

24. Belka, C., Rudner, J., Wesselborg, S., Stepczynska, A., Marini, P., Lepple-Wienhues, A., Faltin, H., Bamberg, M., Budach, W. & Schulze-Osthoff, K. (2000) Differential role of caspase-8 and BID activation during radiation- and CD95-induced apoptosis. *Oncogene* **19**, 1181–1190

25. von Haefen, C., Wieder, T., Essmann, F., Schulze-Osthoff, K., Dörken, B. & Daniel, P.T. (2003) Paclitaxel-induced apoptosis in BJAB cells proceeds via a death receptor-independent, caspase-3/caspase-8-driven mitochondrial amplification loop. *Oncogene* **22**, 2236–2247

26. Radetzki, S., Köhne, C.H., von Haefen, C., Gillissen, B., Sturm, I., Dörken, B. & Daniel, P.T. (2002) The apoptosis promoting Bcl-2 homologues Bak and Nbk/Bik overcome drug resistance in Mdr-1-negative and Mdr-1 overexpressing breast cancer cell lines. *Oncogene* **21**, 227–238

27. Datta, S.R., Katsov, A., Hu, L., Petros, A., Fesik, S.W., Yaffe, M.B. & Greenberg, M.E. (2000) 14-13-3 proteins and survival kinases cooperate to inactivate BAD by BH3 domain phosphorylation. *Mol. Cell* **6**, 41–51

28. Blagosklonny, M.V. (2001) Unwinding the loop of Bcl-2 phosphorylation. *Leukemia* **15**, 869–874

29. Schelwies, K., Sturm, I., Grabowski, P., Scherübl, H., Schindler, I., Hermann, S., Stein, H., Buhr, H.J., Riecken, E.O., Zeitz, M. et al. (2002) Analysis of p53/BAX in primary colorectal carcinoma: low BAX protein expression is a negative prognostic factor in UICC stage III tumors. *Int. J. Cancer* **99**, 589–596

30. Mrozek, A., Petrowsky, H., Sturm, I., Kraus, J., Hermann, S., Hauptmann, S., Lorenz, M., Dörken, B. & Daniel, P.T. (2003) Combined p53/Bax mutation results in extremely poor prognosis in gastric carcinoma with low microsatellite instability. *Cell Death Differ.* **10**, 461–467

31. Güner, D., Sturm, I., Hemmati, P.G., Hermann, S., Hauptmann, S., Wurm, R., Budach, V., Dörken, B., Lorenz, M. & Daniel, P.T. (2003) Multigene analysis of Rb-pathway and apoptosis-control in esophageal squamous cell carcinoma identifies patients with good prognosis. *Int. J. Cancer* **103**, 445–454

32. Prokop, A., Wieder, T., Sturm, I., Essmann, F., Seeger, K., Wuchter, C., Ludwig, W.-D., Henze, G., Dörken, B. & Daniel, P.T. (2000) Relapse in childhood acute lymphoblastic leukemia is associated with decrease of Bax/Bcl-2 ratio and loss of spontaneous caspase-3 processing *in vivo*. *Leukemia* **14**, 1606–1613

33. Baell, J.B. & Huang, D.C. (2002) Prospects for targeting the Bcl-2 family of proteins to develop novel cytotoxic drugs. *Biochem. Pharmacol.* **64**, 851–863

34.   Bosanquet, Sturm, I., Wieder, T., Essmann, F., Bosanquet, M.I., Head, D.J., Dörken, B. & Daniel,
      P.T. (2002) Bax expression correlates with cellular drug sensitivity to doxorubicin, cyclophos-
      phamide and chlorambucil but not fludarabine, cladribine or corticosteroids in B cell chronic
      lymphocytic leukemia. *Leukemia* **16**, 1035–1044

7

Oncogenes as regulators of apoptosis

Mohamed Labazi and Andrew C. Phillips[1]

Medical College of Georgia, Institute of Molecular Medicine and Genetics, CB2803, 1120 15th Street, Augusta, GA 30912, U.S.A.

Abstract

Although cancer is a disease that will afflict one out of three people in the Western world, when considered at a cellular level, it is a rare clonal event. Long-lived organisms, such as humans, have evolved strategies to restrict the development of potentially malignant cells, and one such mechanism is the coupling of proliferative and apoptotic pathways. Multiple oncogenes have the ability to trigger apoptosis when expressed in an inappropriate fashion, and this is thought to restrict tumour formation by eliminating potentially malignant cells that have acquired a mutation stimulating proliferation. Hence for a tumour to arise, in addition to mutations that drive proliferation, mutations that prevent apoptosis are also a prerequisite.

Introduction

Cancer is a disease that is caused by the acquisition of mutations that can modulate the control of growth and survival, and these genetic events affect the expression or function of genes that are broadly classified into two groups, oncogenes and tumour suppppressors. Proto-oncogenes are cellular genes involved in the control of growth and survival under normal physiological conditions, whose activation or over-expression as a consequence of mutation

[1]*To whom correspondence should be addressed (e-mail anphillips@mail.mcg.edu).*

(rendering them oncogenes) can positively contribute to tumour formation. In contrast, tumour suppressors act in normal cells as negative controllers of cell growth and survival, or are involved in the maintenance of genomic integrity and are deleted or are inactive in tumour cells. Consequently, activation of an oncogene is a dominant genetic event, whereas inactivation of a tumour-suppressor gene is normally recessive.

Since cancer is a disease characterized by excessive proliferation, it is not surprising that many oncogenes encode proteins that promote proliferation. The role of proto-oncogenes in regulating normal proliferation has been established since the 1970s, when the first oncogenes were cloned. In contrast, the ability of oncogenes to induce programmed cell death, or apoptosis, was not recognized until the early 1990s. Pioneering work with the c-*myc* oncogene demonstrated that it is also capable of stimulating apoptosis [1]. Expression of the transcription factor c-Myc occurs in normally cycling cells and when ectopically expressed in a quiescent cell is sufficient to drive the cells into S-phase. While activating a proliferative programme, c-Myc also stimulates apoptosis; however, other signals from the external environment that promote survival can suppress this apoptotic programme. Many soluble growth factors can function as survival factors, and are capable of blocking apoptosis in response to c-Myc expression. Under normal physiological conditions, entry of a cell into the cell cycle and the concomitant activation of c-Myc would occur in the presence of these survival signals, preventing c-Myc-induced apoptosis. This proto-oncogene is activated in a number of tumours by over-expression as a consequence of gene amplification or translocation. The outcome of this oncogenic activation of c-Myc, proliferation or apoptosis, is dependent on the presence or absence of survival signals provided by the surrounding environment [2] (Figure 1).

The ability to induce apoptosis has been extended to many other growth-promoting oncogenes using both *in vitro* and *in vivo* systems. One example is E2F1, a member of a family of transcription factors (E2F1–6) that are critical regulators of the cell cycle, mediating expression of a number of genes whose products play regulatory roles in the cell cycle, including the cell-cycle-regulated kinase cdc2, cyclin E and those with a functional role in DNA synthesis itself, such as DNA polymerase α and dihydrofolate reductase [3]. E2F activity is tightly regulated and an important mechanism of control is through interaction with members of the product of the pRB (retinoblastoma susceptibility locus) family. E2F and pRB form a repressive complex that inhibits transcription; however, in response to growth factor signalling, pRB is hyperphosphorylated as cells pass through G_1 and is unable to interact with E2F, enabling these transcription factors to activate expression of genes necessary for DNA synthesis [3]. The important role of this family of transcription factors as regulators of proliferation is reflected in the fact that deregulation of E2F activity is a common event in human cancer and can occur by a number of distinct mechanisms including loss of the tumour suppressor pRB [3].

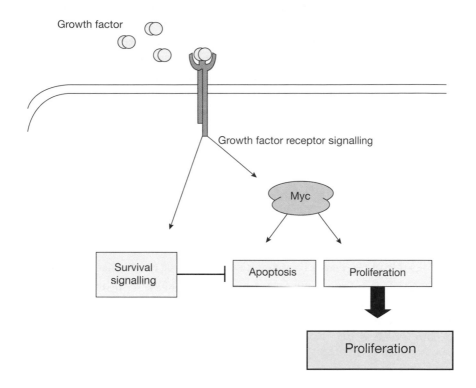

Figure 1. Oncogenes stimulate proliferation and apoptosis
Activation of the proto-oncogene c-*myc* occurs in response to growth factor signalling, stimulating both proliferation and apoptosis. In addition to stimulating c-Myc, growth factors can concomitantly activate survival signalling that blocks the apoptosis stimulated by c-Myc, allowing the cell to proliferate. If c-Myc is activated by an oncogenic event, whether the outcome of this event is proliferation or apoptosis will be determined by the extent of survival signals that the cell receives.

In addition to regulating genes necessary for the cell cycle, forced expression of E2F1 in quiescent cells is sufficient to induce entry into DNA synthesis; however, this is an abortive DNA synthesis, since the cells undergo apoptosis. The ability of E2F1 to induce apoptosis has also been demonstrated *in vivo* in a number of studies. Mice null for pRB fail to develop in part as a result of apoptosis in a number of tissues, illustrating that loss of the normal regulation of E2F is sufficient to trigger apoptosis *in vivo*. Although loss of pRB has consequences in the addition to the deregulation of E2F1 activity, the role of E2F1 in mediating this apoptosis is illustrated in mouse embryos null for both pRB and E2F1, which display much reduced apoptosis, although the embryos still die before birth [4]. A role for E2F1 as a mediator of apoptosis is supported by the analysis of the phenotype of E2F1-null mice; these mice develop normally, but are tumour-prone, exhibiting a reduction in apoptosis in certain tissues [5].

Why do oncogenes stimulate apoptosis? It seems paradoxical that the genes that promote cancer can induce apoptosis. It should be noted that cancer does not arise as a consequence of a single genetic change, rather it requires a number of independent genetic events. In the absence of a compensating mutation blocking apoptosis, activation of a single oncogene may lead to the elimination of the cell by apoptosis, rather than unrestrained proliferation. Thus oncogene-induced apoptosis is a mechanism of tumour suppression, eliminating cells with potentially malignant genetic changes, thus protecting the organism from the development of cancer.

The importance of escape from apoptosis in the development of a tumour was first suggested by the identification and characterization of B-cell lymphoma 2 gene (*bcl-2*). Follicular lymphoma is associated with a reciprocal translocation, which brings the chromosome 18 gene, *bcl-2*, under the regulation of an immunoglobulin locus, deregulating its expression. In these tumours, over-expression of Bcl-2 blocks the normal apoptosis of B-cells, contributing to tumorigenesis by promotion of survival rather than by enhancing proliferation. Bcl-2 is an important regulator of apoptosis, being capable of blocking cell death in response to multiple apoptosis-promoting agents by blocking the activation of caspases, critical proteases that mediate apoptosis [6]. Transgenic studies have demonstrated that the activation of proliferative oncogenes, such as c-Myc, can co-operate with Bcl-2 in tumour formation, underlining the role of apoptosis in restricting tumour development [6].

p53 is a mediator of oncogene-induced apoptosis

What is the molecular mechanism of oncogene-induced apoptosis? There seem to be multiple downstream mediators of this property of oncogenes with a role for both death receptors and mitochondrial apoptotic pathways described. One important mediator of oncogene-induced apoptosis that has received much attention is p53, a critical regulator of apoptosis, and the most commonly mutated gene in human cancer [7].

The transcription factor p53 is activated in response to multiple stresses that may be associated with the initiation or development of cancer, including DNA damage, hypoxia and telomere erosion [2]. This transcription factor can activate mutiple genes that mediate a number of biological programmes, including apoptosis, growth arrest, differentiation and senescence, as well as DNA repair, all of which may contribute to its ability to function as a tumour suppressor.

The DNA tumour viruses, including adenovirus, SV40 (simian virus 40) and human papillomavirus, provide an excellent illustration of the role of p53 in triggering oncogene-induced apoptosis. These small viruses depend on host factors for their replication and so express viral proteins that interfere with the pRB protein family, deregulating E2F activity. This ensures expression of the cellular replicative machinery, but carries with it the drawback of activation of the E2F apoptotic signal. Consequently, each of these distinct viruses encode proteins with the ability to inhibit p53 activity [8]. SV40 large T-antigen and adenovirus E1B 55K bind and inactivate p53, whereas the high-risk human papillomavirus encodes a protein, E6, that binds and mediates the degradation of p53, hence cells expressing these viral oncoproteins are functionally null for p53 (Figure 2). Expression of the viral proteins that deregulate E2F activity in the absence of those that inhibit p53 function converts the outcome from proliferation into apoptosis.

A role for p53 in apoptosis triggered by deregulation of E2F is also supported by a number of *in vivo* studies, including analysis of pRB-null embryos, where loss of p53 function blocks apoptosis in a number of tissues [9]. A role for p53 in oncogene-induced apoptosis is not restricted to the

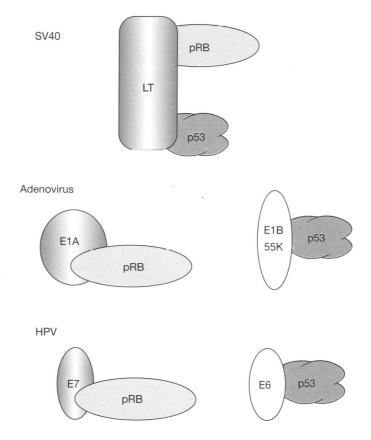

Figure 2. DNA tumour viruses encode proteins that inhibit both pRB family members and p53

The DNA tumour viruses, adenovirus, SV40 and human papillomaviruses (HPV) all encode proteins that inhibit the function of pRB family members thus generating a proliferative signal. Since this also induces an apoptotic response, each of these viruses expresses a protein that inhibits p53 function thus blocking apoptosis. SV40 large T-antigen (LT) and adenovirus E1B 55K bind and inhibit p53 activity. The high-risk HPVs encode a protein, E6, that mediates the ubiquitin-dependent degradation of p53. Thus infection of a cell with any of these viruses renders it functionally null for p53.

deregulation of E2F, rather seems to be a common feature of the proliferation-promoting oncogenes, including c-Myc [2].

ARF stabilizes p53 in response to oncogenic signalling

Oncogenic signalling by over-expression of oncogenes including c-Myc and E2F1 can stabilize p53 [2]. p53 induces transcription of *mdm2*, a gene encoding a ubiquitin ligase that interacts with p53 triggering its ubiquitin-dependent degradation. Hence, under normal conditions p53 is maintained at low levels by this negative-feedback loop; however, in response to stresses that activate p53, such as DNA damage or hypoxia, this feedback loop is broken allowing p53 levels to accumulate (Figure 3).

One mechanism to interrupt this feedback loop is via induction of the ARF (<u>a</u>lternative <u>r</u>eading <u>f</u>rame) of the *INK4a* locus (p19^{ARF} in mouse, p14^{ARF} in humans). This locus remarkably encodes two loci, both of which can function as tumour suppressors, but since they are read in different reading frames, they share no amino acid homology [10]. ARF can stabilize p53 by binding Mdm2 and preventing the degradation of p53. Oncogenes such as Myc, E1a Ras and E2F1 have all been demonstrated to induce expression of this gene,

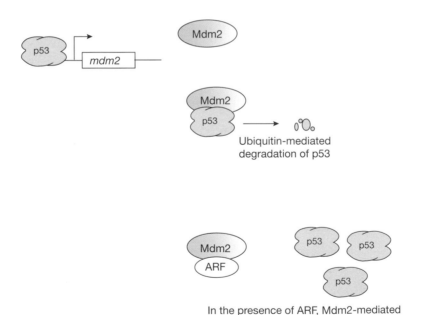

In the presence of ARF, Mdm2-mediated degradation of p53 is blocked

Figure 3. ARF stabilizes p53 by preventing its Mdm2-mediated degradation
Under normal conditions, p53 levels are maintained by a negative-feedback loop; p53 induces the expression of *mdm2*. The product of this gene, Mdm2, binds and mediates the ubiquitin-dependent degradation of p53. Oncogenes can stabilize p53 by expression of ARF, which binds Mdm2, thus preventing the degradation of p53.

leading to stabilization of p53. The mechanism of how diverse oncogenes trigger expression of this gene is unclear; however, evidence suggests that p14^{ARF} may be a direct E2F1 target gene [11].

Almost all tumours contain lesions that lead to deregulation of E2F activity and/or activation of other oncogenes, which would induce ARF, leading to elevated levels of p53. Hence, for a tumour to arise, we would predict that this activation of p53 would have to be abrogated. Whereas loss of p53 would obviously disrupt this pathway, loss of ARF could potentially substitute, and there are experimental data indicating that this can happen. Mice null for ARF develop tumours, indicating that this gene can function as a tumour-suppressor gene, and many tumours that retain wild-type p53 show defects in ARF expression, suggesting that loss of the ability to induce p53 in response to oncogenes allows survival and progression of the tumour cell [12]. This is supported by *in vivo* models using transgenic mice with c-Myc expression targeted to B-cells, where ARF loss could substitute for loss of p53 in the generation of aggressive lymphomas. However, many of the tumours arising in these mice also acquired p53 mutation at a later stage, indicating that there was still selective pressure to lose p53 in the absence of ARF [13].

It is of interest to note that ARF does not appear to be induced upon DNA damage, and loss of ARF does not prevent activation of p53 in response to genotoxic stress, although ARF may contribute to the maintenance of this response. Since certain tumours can tolerate p53 in the absence of ARF, responding to oncogenic signalling may be the critical tumour-suppressing function of p53 in many cell types, rather than the ability to respond to other stresses such as DNA damage [10].

ARF-independent stabilization of p53

While it is clear that ARF can play a role in the stabilization of p53 in response to oncogene expression, emerging evidence indicates that ARF may have p53-independent functions, and additional mechanisms are employed by oncogenes to stabilize p53. As already described, the pRB-null embryo is characterized by apoptosis that is, in part, dependent on p53. In these tissues the elevation of p53 is associated with increased levels of ARF. Surprisingly, in an ARF-null background, the stabilization of p53 is unaffected, indicting that ARF is dispensable for the p53 stabilization seen in response to deregulation of E2F activity [4]. This result is mirrored by other transgenic studies. Targeting of E2F1 expression to the skin leads to induction of ARF, elevation of p53 and apoptosis. In the absence of ARF, the stabilization of p53 is unaffected and the apoptosis is actually increased. In contrast, in the same tissue, c-Myc expression led to the stablization of p53 in an ARF-dependent manner [14]. A similar effect was seen when E2F was deregulated in the brain, with a lack of ARF dependence for the stabilization of p53, despite induction of ARF expression [15].

What underlies these results? It seems clear that, under some circumstances, ARF can promote the stabilization of p53 in response oncogenes, but at least for E2F other mechanisms must exist. It is of interest that ARF has been reported to have p53-independent properties, including the ability to inhibit the cell cycle via a direct interaction with E2F. The extent to which the p53-independent effects of ARF influence its role as a tumour suppressor remains to be determined, but interestingly, triple-knockout mice, null for ARF, p53 and Mdm2, display a greater frequency of tumours than either p53- or p53/mdm2-null mice [16]. While the precise role of ARF in stabilizing p53 in response to oncogene expression is unclear, stabilization of p53 is a critical event in mediating apoptosis regardless of the mechanism of induction.

p53-independent apoptosis

Although p53 is an important mediator of oncogene-induced apoptosis, oncogenes can trigger apoptosis independently of p53. There are now many reports of oncogenes triggering apoptosis in cells null for p53, firmly demonstrating that p53 is not essential for oncogenes to trigger apoptosis. Furthermore, animal models have demonstrated that, *in vivo*, a requirement for p53 to mediate oncogene-induced apoptosis is tissue-specific. In mice null for pRB, although the apoptosis in the central nervous system is p53-dependent, loss of p53 has no impact on the apoptosis in the peripheral nervous system [9].

In addition, although p53 may be essential for apoptosis in certain cell types, even in these cells it is not sufficient. In mouse embryo fibroblasts (MEFs), expression of oncogenes results in stabilization of p53 and apoptosis, whereas, in contrast, expression of p53 alone triggers only a growth arrest [17]. The molecular determinants of whether the outcome of p53 induction is apoptosis or growth arrest are not fully understood. However, recent work has demonstrated that the level of c-Myc can influence the outcome of p53 induction, converting a cell's response from growth arrest to apoptosis. p53 mediates growth arrest by the induction of the cyclin-dependent kinase inhibitor p21, which inhibits the cell cycle, and in certain contexts, expression of this gene can block apoptosis. Myc is recruited to the p21 promoter by the DNA-binding protein Miz and inhibits p53-mediated transcriptional activation of this target gene. In contrast, expression of Myc does not inhibit activation of p53 genes involved in the induction of apoptosis, and in colon cancer cells, this switches the response to p53 expression from growth arrest to apoptosis [18].

Since oncogenes are capable of inducing apoptosis independently of p53 in a number of systems, it seems likely that oncogenes lead to the activation of additional apoptotic pathways that could co-operate with p53 in the induction of apoptosis, in addition to modulating p53 target gene expression. In support of this notion is the identification of a number of genes that are induced by oncogenes that can activate apoptotic pathways.

p73 is a mediator of oncogene-induced apoptosis

Expression of oncogenes can induce the transcriptional activation of the p53 homologue p73, and this related protein can induce growth arrest and apoptosis, sharing many common target genes with p53 [19]. In contrast to oncogene-induced elevation of p53, the induction of p73 is ARF-independent, and, in the case of E2F1, occurs via direct transcriptional activation of the *p73* gene. A role for p73 in oncogene-induced apoptosis has been demonstrated in cell lines using a dominant-negative p73 gene and by the use of compound p53/p73-null mouse cells [20,21].

If p73 is a critical mediator of oncogene-induced apoptosis, we might expect that, like p53, it would function as a tumour-suppressor gene. Although *p73* is not frequently mutated in human tumours, methylation-dependent silencing has been reported in haematological malignancies. Further indirect evidence that inhibition of p73 can contribute to cancer development is provided by the strong bias for expression of p53 mutants that can also inhibit p73 function in human cancers [22]. A role for p73 in tumour development is complicated by the fact that the locus can generate mRNA from two different promoters and the products termed TAp73 and ΔNp73. The ΔNp73 isoforms lack the transactivation domain and, as such, cannot directly induce gene expression, growth arrest or apoptosis. The oligomerization domain of the ΔNp73 is retained and this protein can function as a dominant negative regulator of TAp73 and p53, and is able to inhibit apoptosis induced by both TAp73 and p53. Interestingly, both TAp73 and p53 can induce ΔNp73, thus forming a negative-feedback loop restricting p73 and p53 function. ΔNp73 is thus a potential oncogene, and is over-expressed in a number of tumours, including neuroblastoma, where expression has been demonstrated to be a strong adverse prognostic marker [22].

Is p73 the missing piece of the puzzle explaining how oncogenes induce apoptosis? It is clearly important in certain systems for the ability to trigger apoptosis: compound p53-/p73-null cells are extremely resistant to apoptosis in response to oncogene expression, and MEFs null for both p73 and the related p63 gene are extremely resistant to apoptosis, even though they retain wild-type p53 [21,23]. Interestingly it seems that p53 is unable to induce activation of certain apoptotic target genes [Bax and PERP (p53-activated effector related to PMP-22)] in the absence of its homologues, suggesting co-operation between these family members in the induction of apoptotic target genes [23]. However, as with p53, expression of p73 in MEFs in the absence of oncogenes leads to growth arrest rather than apoptosis. In addition, mutational analysis of E2F1 indicates that apoptosis can be induced by mutants unable to activate p73 [21,24,25]. It seems likely that p53 and p73 individually or in combination activate expression of apoptotic target genes that are necessary, but not sufficient for apoptosis in certain cell types. The oncogenes must trigger additional apoptotic pathways to co-operate with the p53/p73/p63 pathway to trigger apoptosis (Figure 4); this of course begs the question, what are these genes?

Other mediators of oncogene-induced apoptosis

The increasing use of microarray technology is beginning to shed light on the genes activated by oncogenes. In contrast to previous approaches, the problem is too much information. For example, analysis of the effect of over-expressing E2F1 indicates that this results in modulation of the expression of between 1000 and 2000 genes [26]. These E2F-responsive genes encode proteins with diverse functions, potentially contributing to a number of biological programmes, including cell-cycle control, apoptosis, development and differentiation. E2F-responsive genes that may play a role in differentiation and development included members of critical signalling pathways, including the transforming growth factor β superfamily [26]. In fact, even a number of genes encoding products that inhibit apoptosis are induced, including Bcl-2; what role these play in modulating E2F function is unclear. Do small inductions of many genes play a key role in mediating apoptosis or are there a more limited number of critical genes that modulate this process? The examples of ARF and p73 suggest that individual genes can play important roles, but of course this does not eliminate a role for multiple additional genes in the process.

Included in the group of genes that may enhance apoptosis are components of the basic apoptotic machinery itself, Apaf-1 (apoptotic protease-activating factor 1) and the caspases. Caspases are proteases that are key initiators and effectors of the process of apoptosis, cleaving multiple targets leading to the orderly breakup of the cell. In response to the release of cytochrome *c*

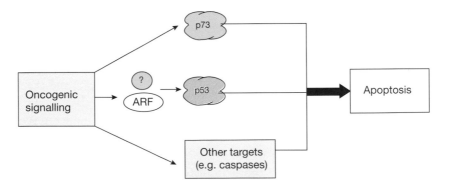

Figure 4. Mechanisms of oncogene-induced apoptosis
Oncogenes can stimulate apoptosis via multiple downstream mediators. Several oncogenes have been demonstrated to stabilize the critical mediator of apoptosis p53, and this stabilization can occur by both ARF-dependent and ARF-independent mechanisms. In addition, oncogenes, including c-Myc and E2F1, can induce the expression of p73, which can play an important role in the induction of apoptosis. In addition to roles for p53 and p73, other target genes are induced that contribute the apoptosis, and it seems likely that a combination of a number of pro-apoptotic mediators co-operates to induce apoptosis in response to oncogenic signalling.

from the mitochondria (an event triggered by a number of apoptotic insults), Apaf-1 processes the pro-caspase-9 to the active form, which then initiates a proleoytic caspase cascade by cleavage of other pro-caspases, leading to the induction of apoptosis. Expression of E1a or E2F1 resulted in significant increases in the levels of both Apaf-1 and several caspases [27,28]. Elevation of components of the basic apoptotic effector machinery would sensitize cells to apoptosis, although signals to trigger activation of the machinery would still be necessary for the induction of apoptosis. The evidence suggests that these genes are direct E2F targets, providing compelling evidence of the intrinsic coupling of replicative and apoptotic pathways.

Oncogenes and cell-intrinsic and cell-extrinsic apoptotic pathways

There are two well-defined apoptotic pathways, termed the cell-extrinsic (death receptor) and the cell-intrinsic (mitochondrial), that can trigger apoptosis. The end point of both of these pathways is the activation of caspases and other molecules involved in mediating apoptosis. The cell-extrinsic pathway triggers apoptosis by the recruitment and activation of the initiator caspase-8 to death receptors in response to ligand, activating the pro-caspase by proteolytic cleavage. This activated initiator caspase can set off a proteolytic cascade of caspase activation, triggering apoptosis. The cell-intrinsic pathway induces apoptosis via the release of pro-apoptotic molecules from the mitochondria, including cytochrome c, which binds to Apaf-1 and caspase-9, triggering activation of caspase-9 and initiating activation of a caspase cleavage cascade.

Cell-intrinsic pathway

The important mediator of oncogene-induced apoptosis p53 is one of a number of signals that has been demonstrated to induce apoptosis by the cell-intrinsic pathway, promoting the release of pro-apoptotic molecules from the mitochondria. The p53 target genes *Bax*, *Puma* and *Noxa* encode important mediators of p53-induced apoptosis, and function as positive promoters of the cell-intrinsic pathway [7]. The most persuasive evidence for a role of the mitochondrial pathway in oncogene-induced apoptosis is that provided by analysis of cells from mice null for various components of the apoptotic machinery. Mice deficient in either Apaf-1 or caspase-9, but not caspase-8, showed a severely impaired apoptotic response to oncogenic signalling, suggesting that oncogenes trigger apoptosis via the cell-intrinsic pathway. These investigators extended their results with transformation assays, indicating that loss of Apaf-1 or caspase-9 could function, instead of loss of p53, and co-operate with proliferative oncogenes in tumour formation, suggesting that these genes could function as tumour suppressors [29]. This hypothesis is supported by analysis of melanomas, in which *apaf-1* is often silenced by methylation [30].

These data firmly establish a role for the cell-intrinsic pathways in mediating oncogene-induced apoptosis in certain cell types.

Cell-extrinsic pathway

A wealth of data also implicates a role for mitochondrion-independent pathways in apoptosis triggered by oncogenes. As already discussed, p53 is a critical mediator of oncogene-induced apoptosis, transcribing genes that function in the mitochondrial pathway. However, p53 also induces the expression of genes involved in death-receptor signalling, including the death receptors *dr5* and *fas*; these receptors are capable of activating caspase in response to exposure to the appropriate ligand. In addition to inducing expression of genes involved in death-receptor signalling, p53 can also increase trafficking of these receptors to the cell surface [31]. Whether any of these targets would trigger apoptosis in a cell is likely to be dependent on a number of different signals, not least of which would be the presence of the death-receptor ligand.

E1a, Myc and E2F can all sensitize cells to the presence of certain death-receptor ligands, such as TNFα (tumour necrosis factor α), and in the case of c-Myc, death-receptor signalling has been shown to be necessary for apoptosis in certain cells [2]. One such ligand, TRAIL (TNF-related apoptotis-inducing ligand)/Apo2L shows a remarkable ability to induce apoptosis specifically in tumour cells, and is currently undergoing clinical trails as an anti-cancer agent. The molecular basis of this sensitivity of tumour compared with normal cells is unclear, but it does not seem to be determined by the presence of the receptors. Whatever the mechanism, it seems that the oncogenes can rewire cells to become sensitive to apoptosis in response to this ligand. In this light, it is interesting that both Myc and E2F1 have been shown to modulate the signalling from the TNF receptor. In addition to activating caspases, the TNF receptor can trigger additional signal-transduction pathways that inhibit apoptosis, including activation of the transcription factor nuclear factor κB. Nuclear factor κB is an important promoter of survival, inducing the expression of genes that inhibit apoptosis. Inhibition of nuclear factor κB activity has been demonstrated to sensitize cells to apoptosis in response to multiple inducers of apoptosis. Expression of Myc and E2F1 can block activation of this anti-apoptotic signalling without interfering with the activation of caspase 8, rendering the cells sensitive to TNFα [32,33]. The target genes of these transcription factors that are responsible for this phenotype remain to be identified.

There is also *in vivo* evidence that death receptors may play a role in oncogene-induced apoptosis. Generation of pRB-null embryos in a variety of genetic backgrounds has been informative in determining the mediators of apoptosis in response to deregulated E2F. Analysis of compound pRB-/Apaf-1-null embryos indicated that the apoptosis generated in certain tissues is not dependent on Apaf-1, eliminating an obligatory role for the cell-intrinsic pathway [34]. Of course, this does not prove that the apoptosis is the result of

death-receptor signalling, but demonstrates that apoptosis generated by deregulation of E2F is mediated by distinct apoptotic pathways in a tissue-specific
fashion.

If the death-receptor pathway is a critical mediator of oncogene-induced
apoptosis, we would expect mutations that abrogate the activity of this pathway in human cancer. One such case appears to be neuroblastoma, in which
mutation or loss of expression of caspase-8 appears to be extremely common
in cases where there is amplification of N-Myc [35]. Since N-Myc can induce
apoptosis, it may be that the loss of caspase-8 allows the tumour to tolerate the
abnormally high levels of this pro-apoptotic oncogene.

Different oncogenes, same mechanisms?

A diverse array of proliferative oncogenes can trigger apoptosis, so are there
common features, or a distinct series of mechanisms employed by different
oncogenes? A definitive answer cannot yet be given, but certain features seem
to be common to a number of oncogenes: several oncogenes have been
demonstrated to induce apoptosis associated with a stabilization of p53, ARF
can be induced by a number of different oncogenes, and p73 plays a role in
apoptosis by several oncogenes.

What is interesting is that both ARF and p73 (and the caspases) are direct
E2F target genes, suggesting that perhaps oncogenes could function via a stimulation of free E2F, driving both proliferation and apoptotic pathways
[11,20,21,27]. Despite E2F1 and E2F2 being capable of inducing entry into
DNA synthesis, mice null for these transcription factors develop normally, but
are tumour-prone [5]. This has led to the suggestion that certain E2Fs may function to detect aberrant or persistent proliferation and activate an apoptotic programme. Whether E2Fs play a role in mediating apoptosis in response to other
oncogenes or whether there are distinct mechanisms remains to be determined.

Conclusions

Excessive mitotic signalling by activated oncogenes is coupled to apoptosis as a
failsafe mechanism that prevents oncogenic transformation. For a fully
malignant cell to emerge, this apoptosis must be overcome. This is reflected in
the co-operation between genetic events that stimulate proliferation and those
that inhibit apoptosis in the development of cancer.

This coupling of proliferative and apoptotic pathways provides a window
of opportunity for cancer researchers. During the development of a tumour,
compensating mutations, for example loss of p53 or ARF, may arise, giving
protection against apoptosis stimulated by activation of oncogenes.
Nevertheless, although tumour cells have to a certain extent escaped this apoptotic mechanism, some tumour cells still show enhanced sensitivity to apoptosis, a fact that may underlie the efficacy of many chemotherapeutic protocols.
The challenge for investigators is to develop rational therapies to exploit the

apoptotic pathways stimulated by oncogenic activation, or reactivate the apoptotic pathways that are lost during the development of tumours.

Summary

- *Expression of oncogenes can trigger apoptosis in primary cells. This property protects the organism from oncogenic change; mutation leading to the activation of growth promoting oncogenes can result in the cell being eliminated by apoptosis.*

- *Since oncogene-induced apoptosis acts as a brake on tumour development, in addition to mutations that stimulate proliferation, mutations that inhibit apoptosis are a prerequisite for tumour formation.*

- *The tumour suppressor p53 is a critical mediator of oncogene-induced apoptosis in a number of tissues. Oncogenes can stabilize p53, by both ARF-dependent and ARF-independent mechanisms.*

- *Oncogenes can trigger apoptosis independently of p53, via a number of different downstream mediators including the p53 homologue p73.*

A.C.P. is a Georgia Cancer Coalition Distinguished Scholar.

References

1. Evan, G.I., Wyllie, A.H., Gilbert, C.S., Littlewood, T.D., Land, H., Brooks, M., Waters, C.M., Penn, L.Z. & Hancock, D.C. (1992) Induction of apoptosis in fibroblasts by c-Myc protein. *Cell* **69**, 119–128

2. Evan, G.I. & Vousden, K.H. (2001) Proliferation, cell cycle and apoptosis in cancer. *Nature (London)* **411**, 342–348

3. Nevins, J.R. (2001) The Rb/E2F pathway and cancer. *Hum. Mol. Genet.* **10**, 699–703

4. Tsai, K.Y., Hu, Y., Macleod, K.F., Crowley, D., Yamasaki, L. & Jacks, T. (1998) Mutation of E2f-1 suppresses apoptosis and inappropriate S phase entry and extends survival of Rb-deficient mouse embryos. *Mol. Cell* **2**, 293–304

5. DeGregori, J. (2002) The genetics of the E2F family of transcription factors: shared functions and unique roles. *Biochim. Biophys. Acta Rev. Cancer* **1602**, 131–150

6. Cory, S. & Adams, J.M. (2002) The Bcl2 family: regulators of the cellular life-or-death switch. *Nat. Rev. Cancer* **2**, 647–656

7. Vousden, K.H. & Lu, X. (2002) Live or let die: the cell's response to p53. *Nat. Rev. Cancer* **2**, 594–604

8. Roulston, A., Marcellus, R.C. & Branton, P.E. (1999) Viruses and apoptosis. *Annu. Rev. Microbiol.* **53**, 577–628

9. Macleod, K. (1999) pRb and E2f-1 in mouse development and tumorigenesis. *Curr. Opin. Genet. Dev.* **9**, 31–39

10. Sherr, C.J. (2001) The INK4a/ARF network in tumour suppression. *Nat. Rev. Mol. Cell Biol.* **2**, 731–737

11. Bates, S., Phillips, A.C., Clark, P.A., Stott, F., Peters, G., Ludwig, R.L. & Vousden, K.H. (1998) p14(ARF) links the tumour suppressors RB and p53. *Nature (London)* **395**, 124–125

12. Kamijo, T., Zindy, F., Roussel, M.F., Quelle, D.E., Downing, J.R., Ashmun, R.A., Grosveld, G. & Sherr, C.J. (1997) Tumor suppression at the mouse INK4a locus mediated by the alternative reading frame product p19ARF. *Cell* **91**, 649–659

13. Schmitt, C.A., McCurrach, M.E., de Stanchina, E., Wallace-Brodeur, R.R. & Lowe, S.W. (1999) INK4a/ARF mutations accelerate lymphomagenesis and promote chemoresistance by disabling p53. *Genes Dev.* **13**, 2670–2677

14. Russell, J.L., Powers, J.T., Rounbehler, R.J., Rogers, P.M., Conti, C.J. & Johnson, D.G. (2002) ARF differentially modulates apoptosis induced by E2F1 and Myc. *Mol. Cell Biol.* **22**, 1360–1368

15. Tolbert, D., Lu, X., Yin, C., Tantama, M. & Van Dyke, T. (2002) p19(ARF) is dispensable for oncogenic stress-induced p53-mediated apoptosis and tumor suppression *in vivo. Mol. Cell Biol.* **22**, 370–377

16. Weber, J.D., Jeffers, J.R., Rehg, J.E., Randle, D.H., Lozano, G., Roussel, M.F., Sherr, C.J. & Zambetti, G.P. (2000) p53-independent functions of the p19(ARF) tumor suppressor. *Genes Dev.* **14**, 2358–2365

17. Zindy, F., Eischen, C.M., Randle, D.H., Kamijo, T., Cleveland, J.L., Sherr, C.J. & Roussel, M.F. (1998) Myc signaling via the ARF tumor suppressor regulates p53-dependent apoptosis and immortalization. *Genes Dev.* **12**, 2424–2433

18. Seoane, J., Le, H.V. & Massague, J. (2002) Myc suppression of the p21(Cip1) Cdk inhibitor influences the outcome of the p53 response to DNA damage. *Nature (London)* **419**, 729–734

19. Zaika, A., Irwin, M., Sansome, C. & Moll, U.M. (2001) Oncogenes induce and activate endogenous p73 protein. *J. Biol. Chem.* **276**, 11310–11316

20. Stiewe, T. & Putzer, B.M. (2000) Role of the p53-homologue p73 in E2F1-induced apoptosis. *Nat. Genet.* **26**, 464–469

21. Irwin, M., Marin, M.C., Phillips, A.C., Seelan, R.S., Smith, D.I., Liu, W.G., Flores, E.R., Tsai, K.Y., Jacks, T., Vousden, K.H. & Kaelin, W.G. (2000) Role for the p53 homologue p73 in E2F-1-induced apoptosis. *Nature (London)* **407**, 645–648

22. Melino, G., De Laurenzi, V. & Vousden, K.H. (2002) p73: friend or foe in tumorigenesis. *Nat. Rev. Cancer* **2**, 605–615

23. Flores, E.R., Tsai, K.Y., Crowley, D., Sengupta, S., Yang, A., McKeon, F. & Jacks, T. (2002) p63 and p73 are required for p53-dependent apoptosis in response to DNA damage. *Nature (London)* **416**, 560–564

24. Phillips, A.C., Bates, S., Ryan, K.M., Helin, K. & Vousden, K.H. (1997) Induction of DNA synthesis and apoptosis are separable functions of E2F-1. *Genes Dev.* **11**, 1853–1863

25. Hsieh, J.K., Fredersdorf, S., Kouzarides, T., Martin, K. & Lu, X. (1997) E2F1-induced apoptosis requires DNA binding but not transactivation and is inhibited by the retinoblastoma protein through direct interaction. *Genes Dev.* **11**, 1840–1852

26. Muller, H., Bracken, A.P., Vernell, R., Moroni, M.C., Christians, F., Grassilli, E., Prosperini, E., Vigo, E., Oliner, J.D. & Helin, K. (2001) E2Fs regulate the expression of genes involved in differentiation, development, proliferation, and apoptosis. *Genes Dev.* **15**, 267–285

27. Nahle, Z., Polakoff, J., Davuluri, R.V., McCurrach, M.E., Jacobson, M.D., Narita, M., Zhang, M.Q., Lazebnik, Y., Bar-Sagi, D. & Lowe, S.W. (2002) Direct coupling of the cell cycle and cell death machinery by E2F. *Nat. Cell Biol.* **4**, 859–864

28. Moroni, M.C., Hickman, E.S., Denchi, E.L., Caprara, G., Colli, E., Cecconi, F., Muller, H. & Helin, K. (2001) Apaf-1 is a transcriptional target for E2F and p53. *Nat. Cell Biol.* **3**, 552–558

29. Soengas, M.S., Alarcon, R.M., Yoshida, H., Giaccia, A.J., Hakem, R., Mak, T.W. & Lowe, S.W. (1999) Apaf-1 and caspase-9 in p53-dependent apoptosis and tumor inhibition. *Science* **284**, 156–159

30. Soengas, M.S., Capodieci, P., Polsky, D., Mora, J., Esteller, M., Opitz-Araya, X., McCombie, R., Herman, J.G., Gerald, W.L., Lazebnik, Y.A. et al. (2001) Inactivation of the apoptosis effector Apaf-1 in malignant melanoma. *Nature (London)* **409**, 207–211

31. Vousden, K.H. (2000) p53: death star. *Cell* **103**, 691–694

32. Phillips, A.C., Ernst, M.K., Bates, S., Rice, N.R. & Vousden, K.H. (1999) E2F-1 potentiates cell death by blocking antiapoptotic signaling pathways. *Mol. Cell* **4**, 771–781

33. Klefstrom, J., Arighi, E., Littlewood, T., Jaattela, M., Saksela, E., Evan, G.I. & Alitalo, K. (1997) Induction of TNF-sensitive cellular phenotype by c-myc involves p53 and impaired NF-kappa B activation. *EMBO J.* **16**, 7382–7392

34. Guo, Z., Yikang, S., Yoshida, H., Mak, T.W. & Zacksenhaus, E. (2001) Inactivation of the retinoblastoma tumor suppressor induces apoptosis protease-activating factor-1 dependent and independent apoptotic pathways during embryogenesis. *Cancer Res.* **61**, 8395–8400

35. Teitz, T., Wei, T., Valentine, M.B., Vanin, E.F., Grenet, J., Valentine, V.A., Behm, F.G., Look, A.T., Lahti, J.M. & Kidd, V.J. (2000) Caspase 8 is deleted or silenced preferentially in childhood neuroblastomas with amplification of MYCN. *Nat Med.* **6**, 529–535

8

The final step in programmed cell death: phagocytes carry apoptotic cells to the grave

Aimee M. deCathelineau and Peter M. Henson[1]

Program in Cell Biology, Department of Pediatrics, National Jewish Medical and Research Center, 1400 Jackson Street, Denver, CO 80206, U.S.A.

Abstract

As cells undergo apoptosis, they are recognized and removed from the body by phagocytes. This oft-overlooked yet critical final step in the cell-death programme protects tissues from exposure to the toxic contents of dying cells and also serves to prevent further tissue damage by stimulating production of anti-inflammatory cytokines and chemokines. The clearance of apoptotic-cell corpses occurs throughout the lifespan of multicellular organisms and is important for normal development during embryogenesis, the maintenance of normal tissue integrity and function, and the resolution of inflammation. Many of the signal-transduction molecules implicated in the phagocytosis of apoptotic cells appear to have a high degree of evolutionary conservation, and therefore the engulfment of apoptotic cells is likely to represent one of the most primitive forms of phagocytosis. With the realization that the signals that govern apoptotic-cell removal also serve to attenuate inflammation and the immune response, as well as initiate signals for tissue repair and remodelling in response to cell death, the study of apoptotic cell clearance is a field experiencing a dynamic increase in interest and momentum.

[1]*To whom correspondence should be addressed (e-mail hensonp@njc.org).*

Introduction

Strict control of the cell-death programme is important for normal deletion of unwanted cells, and aberrations in this process have obvious and profound implications for the pathogenesis of disease. However, the energy invested in apoptosis is arguably for naught in the absence of efficient cell corpse removal. The contents of dying cells are a potent primordial 'biohazard', with potential to inflict extensive tissue damage as a result of cytotoxicity, invocation of inappropriate inflammatory responses and presentation of auto-antigens [1–3]. Indeed, the failure to remove apoptotic cells efficiently has been linked to the pathogenesis of a variety of different chronic inflammatory and auto-immune diseases, including systemic lupus erythematosus and retinitis pigmentosa [4,5]. Therefore it is essential that dying cells are recognized and removed with immunological stealth to maintain normal tissue homoeostasis in the face of cell death.

Apoptotic-cell clearance and the inflammatory response initiated by the phagocyte–corpse interaction is a multi-faceted field of study which is complicated by the simultaneous presence of many different putative ligands on the apoptotic-cell surface, and a wide variety of potential receptors expressed alone or in combination by many different types of phagocyte. In addition, it is becoming increasingly clear that redundant pathways for engulfment exist, and that the cytokine environment in which the engulfment process occurs also plays an important role in the regulation of apoptotic-cell uptake and subsequent immunological reaction by the host. While much has been learned in the past 20 years regarding the phagocytosis of apoptotic cells (for comprehensive reviews, see [1–3,6–8]), many aspects of this intricate process remain a mystery.

ACAMPs (apoptotic-cell-associated molecular patterns): cell-surface changes make apoptotic cells appetizing to phagocytes

How apoptotic cells are detected and targeted for removal by phagocytes in the context of normal, healthy tissue is a central issue in the study of apoptotic-cell removal. Phagocytosis of micro-organisms is mediated by recognition of 'PAMPs', or pathogen-associated molecular patterns, on bacterial outer membranes through the innate immune system. Similarly, the surface changes that promote apoptotic-cell removal have come to be known as ACAMPs. A variety of factors, including the tissue type of the apoptotic cell and the stimulus that initiates apoptosis, could conceivably affect ACAMP expression. In addition, apoptotic cells are not static entities and ACAMPs can reasonably be expected to change during progression of the cell-death programme [9].

Several different types of ACAMPs have been identified [2,6,7] (Figure 1). At an early stage in the cell-death programme, cells lose the ability to maintain phospholipid asymmetry, and as a result, expose PS (phosphatidylserine) on the outer leaflet of the plasma membrane [10]. In addition, proteins on the apoptotic-cell surface may display altered glycosylation patterns, such as increased expression of N-acetylglucosamine, which is thought to promote phagocytosis through lectin–carbohydrate interactions [11]. Finally, cell-surface proteins such as ICAM-3 (intercellular cell-adhesion molecule 3) [12] may

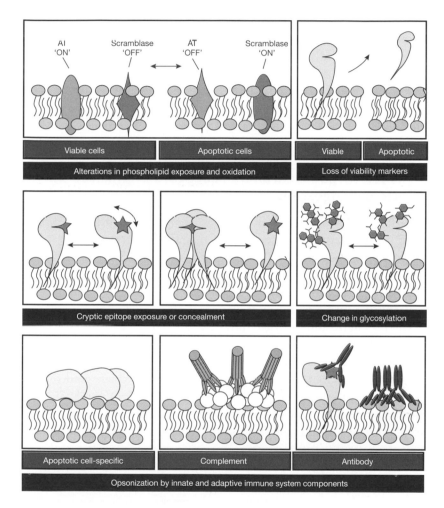

Figure 1. ACAMP expression designates apoptotic cells for phagocytosis

As cells undergo apoptosis, proteins and lipids on the cell surface become altered as result of the cell-death programme. These molecular patterns are important for recognition and removal of the apoptotic cell by phagocytes. Identified ACAMPs include loss of phospholipid asymmetry and PS (phosphatidylserine) exposure, oxidation of phospholipids (top left), exposure of cryptic epitopes as a result of changes in protein conformation or protein–protein interactions (centre left), and alterations in glycosylation patterns (centre right). In addition to gain of specific molecular patterns, signals which prevent phagocytosis and indicate viability of normal cells ('VCAMPs', viable-cell-associated molecular patterns) are lost from apoptotic cells (top right). Finally, apoptotic cells can become opsonized by molecules that recognize PS, such as MFG-E8 (lactadherin), Gas-6 or Protein S (bottom left), by components of the innate immune system, including complement, pentraxin and collectin family members (bottom centre), or by apoptotic-cell-specific autoantibodies (bottom right). AT, aminophospholipid translocase; MFG-E8, milk fat globule-EGF-factor 8; EGF, epithelial growth factor.

also undergo changes in conformation, associate with other proteins or move into alternative lipid microenvironments, resulting in the presentation of epitopes which are normally masked on viable cells.

In contrast to epitopes and engulfment signals that are gained during apoptosis, dying cells may also lose expression of molecules that are normally present on the surface of viable cells. Such 'don't eat me' signals have long been speculated to exist, and CD31 [PECAM-1 (platelet–endothelial cell adhesion molecule-1)] has recently been demonstrated to prevent binding of viable lymphocytes to endothelial cell surfaces, thus preventing inadvertent phagocytosis [13]. Although this particular (and first) example is relatively cell-type-specific, 'VCAMP' (viable-cell-associated molecular pattern) expression may represent yet another level in the complex series of surface pattern changes which regulate apoptotic cell recognition.

Utensils for corpse clearance: receptors implicated in apoptotic-cell ingestion

Many different phagocytic receptors have been implicated in apoptotic-cell recognition, primarily by antibody and ligand-inhibition studies, but also through the use of genetic manipulation in *Caenorhabditis elegans*, *Drosophila melanogaster* and mice (Figure 2). Redundant pathways apparently mediate phagocytosis of apoptotic cells, as treatment of phagocytes with any given receptor-specific antibody or ligand often will not inhibit uptake by more than 50%. Even genetic deletion of individual receptors rarely results in a significant reduction of apoptotic-cell clearance *in vivo*. It is important to note that recognition systems employed by the phagocyte may vary according to phagocyte type and maturation stage, and that both primary and secondary mechanisms for apoptotic-cell removal may exist [14]. Whereas immediate recognition by neighbouring phagocytes may be mediated by early changes on the cell surface, opsonization with apoptotic-cell-specific binding proteins may enhance clearance in situations where apoptosis is occurring at a high rate. A failure to clear apoptotic cells efficiently may result in secondary opsonization with complement or antibody, and thus increase the likelihood of inappropriate inflammatory responses and contribute to the pathogenesis of autoimmune disease.

A number of receptors, including scavenger receptor family members {CD36, CD68, SRA (scavenger receptor A) and others [15]}, the glycosylphosphatidylinositol-linked lipopolysaccharide receptor CD14 [16] and the PSR (PS receptor) [17], are thought to be capable of direct ACAMP recognition. In addition, several molecules that recognize PS are thought to bridge the apoptotic cell to the phagocyte, essentially acting as apoptotic-cell opsonins. These include the soluble growth factor Gas-6, which is thought to promote phagocytosis by binding to the Tyro-3 receptor tyrosine kinase family member Mer [4,18]. In addition, thrombospondin [19] and MFG-E8 (lactadherin; milk fat globule-epithelial growth factor-factor 8) [20] are thought to form complexes with CD36 and $\alpha_v\beta_3$ integrin (CD51/61). The serum component Protein S was also recently shown to enhance phagocytosis of apoptotic cells [21], although the receptor that mediates this interaction on the phagocyte has not yet been identified.

Apoptotic cells may also be opsonized by classic components of both the innate and adaptive immune system. The pentraxin family of acute-phase proteins have been shown to enhance phagocytosis of apoptotic cells while maintaining an anti-inflammatory response [22]. The complement component C1q, and other members of the collectin family, including mannose-binding lectin and lung surfactant proteins A and D, have also been shown to promote phagocytosis of apoptotic cells through interaction with surface-expressed calreticulin, which is in turn bound to CD91 (LRP-1, $\alpha 2$ macroglobulin receptor) [23]. Finally, disturbances in normal apoptotic-cell clearance may lead to the inappropriate generation of apoptotic-cell-specific antibodies, which are found with high frequency in patients with auto-immune disease [5].

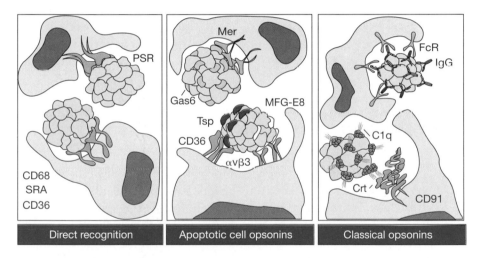

Figure 2. Engulfment of apoptotic cells may be mediated by many different receptor-mediated recognition systems

A variety of receptors and receptor complexes have been demonstrated to be involved in the regulation of apoptotic-cell clearance. These include receptors capable of direct recognition of surface changes on the apoptotic cell (left-hand panel), such as the glycosylphosphatidylinositol-linked lipopolysaccharide receptor CD14 (ICAM-3), the PSR (PS receptor) and members of the scavenger receptor superfamily, including the oxidized low-density lipoprotein receptor, CD68, SRA and CD36. Other receptors are thought to mediate engulfment by binding of 'bridging molecules', which specifically bind to the apoptotic cells (centre panel). Examples of such systems include binding of the soluble factor Gas-6 by the Mer receptor tyrosine kinase, and a complex consisting of CD36, $\alpha_v\beta_3$ integrin (CD51/61) and thrombospondin, or $\alpha_v\beta_3$ integrin in conjunction with MFG-E8 (lactadherin). Finally, phagocytosis of apoptotic cells may also be facilitated through opsonization with components of the innate and adaptive immune system (right-hand panel). The complement component C1q is thought to bind apoptotic cells through its globular head domain, and binds to surface-bound calreticulin, which is complexed to CD91, via its collagenous tail domain. Apoptotic cells opsonized with auto-antibodies may be phagocytosed through the Fc family of receptors (FcR).

Tether and tickle: PS can stimulate phagocytosis of adherent particles

Because phagocytes often express more than one type of apoptotic-cell receptor, deciphering how each receptor contributes to the engulfment process has been difficult. Recent studies using an erythrocyte-based system to target specific, individual receptors on the phagocyte suggest that many of the receptors proposed to be involved in apoptotic-cell clearance mediate binding of the apoptotic cell to the phagocyte, whereas the presence of PS was required for particle internalization [24]. Addition of PS-containing microliposomes or antibody-mediated ligation of the PSR could even drive the internalization of particles bound to the cell surface, which were otherwise not engulfed. These results led to the generation of the 'tether and tickle' hypothesis, which suggests that phagocytosis of apoptotic cells may require a binding step that may often be separate from a second, ligand-induced, stimulation of the uptake process. In this model, specificity for apoptotic-cell uptake and the subsequent phagocyte response could conceptually be transmitted through either interaction. Furthermore, 'tether and tickle' does not imply that the binding event is entirely passive in terms of phagocytosis, nor does it exclude the possibility that recognition of apoptotic cells by specific receptors or receptor complexes may modulate the ability of the cell to respond to the engulfment signal.

'Efferocytosis': uptake of apoptotic cells occurs through a unique phagocytic mechanism

Although only a few cell types express specific receptors for classical phagocytic opsonins, such as immunoglobulins, most eukaryotic cells do have the latent and very primitive ability to phagocytose particulate matter from their extracellular environment, including apoptotic cells and apoptotic bodies. In fact, the task of removing apoptotic cells is conducted by cells of nearly all tissue types, including fibroblasts, endothelial cells and epithelial cells, and is a rapid process *in vivo*, such that apoptotic-cell bodies are rarely observed outside of other cells *in situ*. Clearance of apoptotic cells, although delayed, even occurs in the PU.1-knockout mouse, which is macrophage-deficient [25]. Therefore, it is predicted that the mechanism for apoptotic-cell engulfment, including the elements recognized on the apoptotic body, the receptors involved in recognition and uptake, and the signal-transduction pathways utilized during engulfment, may be evolutionarily conserved among species and expressed ubiquitously, allowing a wide variety of cell types to participate in the clearance process.

Phagocytosis, defined as engulfment of particles larger than 0.5 μm in diameter, can occur through a variety of mechanisms that are distinguished by both morphological characteristics and by the molecular pathways which regulate them [26]. Phagocytosis of IgG-opsonized particles is mediated by a 'zipper' mechanism, wherein pseudopod extension is driven by sequential binding of ligands on the particle being ingested to receptors on the phagocyte, resulting in the formation of a phagosome with a membrane that is very tightly opposed to the ingested particle or cell. Phagocytosis mediated by complement is described as a 'sinking' mechanism, as the particle being engulfed appears to settle into the body of the phagocyte with very little membrane protrusion. Finally, the process of macropinocytosis can entrap particles or cells bound to the surface of the phagocyte when the ruffling membrane folds back against the cell body and fuses [27]. Macropinocytosis regulates bulk fluid and macromolecule uptake from the extracellular space, and is the mechanism utilized by immature dendritic cells for antigen acquisition. In contrast to the zipper mechanism, phagosomes formed as a result of macropinocytosis are loosely opposed to the ingested particle or cell, as typified by the 'spacious phagosomes' formed during invasion of *Salmonella*. All described mechanisms of phagocytosis require actin cytoskeletal rearrangement, and thus are regulated by the Rho family of low-molecular-mass GTPases. While there is some controversy within the literature regarding how these molecules regulate different forms of phagocytosis, Cdc42 and Rac appear to be important for both FcR (Fc receptor)-mediated phagocytosis and macropinocytosis, while RhoA plays a dominant role in regulation of complement-receptor-mediated phagocytosis (CR3).

Phagocytosis of apoptotic cells appears to resemble a process similar to macropinocytosis, rather than a classical zipper mechanism, although these two means for engulfment of debris, cell bodies and pathogens share a requirement for the activation of a number of different signal-transduction molecules, including Cdc42, Rac1 and phosphoinositide 3-kinase [28]. Apoptotic cells, in contrast to IgG-opsonized viable cells, are found enclosed within spacious phagosomes (A. deCathelineau, unpublished work). In addition, similar to macropinocytosis, engulfment of apoptotic cells is sensitive to amiloride, whereas phagocytosis of IgG-opsonized particles is not [24,27]. These observations, combined with the observation that PS can stimulate membrane ruffling and the formation of macropinosomes, as well as initiate the phenomenon of 'bystander engulfment', distinguish further the engulfment of apoptotic cells as a unique phagocytic mechanism [24]. To emphasize the idea that apoptotic-cell clearance is a unique form of phagocytosis with important physiological consequences for the regulation of inflammation, the members of this laboratory have begun to use the term 'efferocytosis', derived from the Latin prefix *effero-*, meaning 'to take away, to put away, to carry to the grave, or to bury', to describe recognition and engulfment of apoptotic cells.

Dinner conversations: is there an apoptotic-cell-engulfment synapse?

Recently, Grimmer and colleagues [29] demonstrated that cholesterol was important for the regulation of macropinocytosis, and recent studies have shown that cholesterol is also important for the phagocytosis of apoptotic cells, but not for FcR-mediated engulfment (A.C. Tosello-Trampont, K.S. Ravichandran, A.M. deCathelineau and P.M. Henson, unpublished work). During phagocytosis of apoptotic cells, the diffuse distribution of the ganglioside G_{M1}, a lipid raft marker, appears to be altered into a large aggregated patch at the interface of the apoptotic cell and the phagocyte (A.C. Tosello-Trampont, K.S. Ravichandran, A.M. deCathelineau and P.M. Henson, unpublished work). These observations support further the concept that the mechanism that governs apoptotic-cell removal is distinct from other mechanisms of phagocytosis, and that a redistribution of lipid microdomains and proteins on the surface of the phagocyte may occur during the engulfment process. The concept of an apoptotic-cell 'synapse', such as that observed during antigen presentation and T-cell activation, has been suggested previously [2,14]; however, the evidence for the existence of such a structure in this context is only beginning to emerge.

The activation state of some signalling molecules is known to be effected by their lipid environment, and lipid rafts are thought to act as platforms for the amplification of signal-transduction pathways. Several of the putative receptors for apoptotic-cell clearance are known to be enriched within cholesterol- and sphingolipid-dense lipid raft domains, including $\alpha_v\beta_3$ integrin and the glycosylphosphatidylinositol-linked protein CD14. The adaptor protein CrkII and Dock180, as well as the Rho-family GTPase Rac1, are recruited to the engulfment site by integrins during uptake of apoptotic cells [30]. It is interesting to speculate that efferocytosis may be a hybrid mechanism between macropinocytosis and the specialized zipper mechanism. While stimulated macropinocytosis can form vesicles as large as 2.0 μm in diameter and mediate the engulfment of fluid, macromolecules and micro-organisms, it is probably not capable of entrapping whole apoptotic-cell bodies. Therefore, the aggregation of lipid rafts at the interface of the apoptotic cell and the phagocyte during efferocytosis may serve to recruit and accumulate apoptotic-cell-receptor complexes and signalling molecules to promote an emphatic signalling burst and drive particle engulfment, whereas extension of membrane during phagocytic 'zippering', such as that mediated by the FcR, is localized and driven by sequential ligand–receptor-binding events (Figure 3).

Cell-engulfment defects: genetic studies in *C. elegans* define signal-transduction molecules required for efficient phagocytosis of apoptotic cells

Much of what is known about the signal-transduction pathways that regulate efferocytosis comes from genetic studies in the nematode *C. elegans* [6,8]. Although *C. elegans* has no macrophages, or even an immune system, each of the 131 cells that die during the development of the hermaphrodite worm is swiftly recognized and engulfed, as they are in higher organisms. The study of

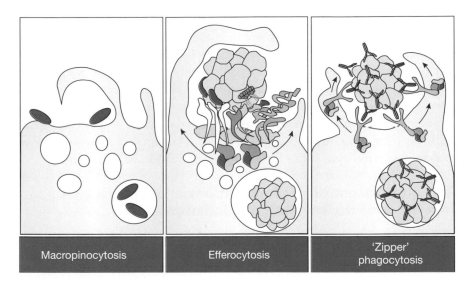

Figure 3. Efferocytosis is a distinct mechanism of phagocytosis

Macropinocytosis (left-hand panel) may be stimulated by growth factors or may occur constitutively in some cell types, such as immature dendritic cells. This mechanism of uptake results in the uptake of fluid, macromolecules and micro-organisms into heterogeneous vesicles, 0.2–2.0 μm in diameter. While PS can stimulate macropinosome formation, engulfment of apoptotic cells (centre panel) stimulates a reorganization of lipid raft microdomains to the interface of the apoptotic cell and the phagocyte, and probably requires the assembly of a supermolecular complex for amplification of signal-transduction pathways and uptake of the cell body. In contrast, phagocytosis mediated by a zipper mechanism (right-hand panel), as typified by phagocytosis of IgG-opsonized particles by the FcR, drives pseudopod extension and uptake through sequential and localized activation of signal-transduction pathways through ligand–receptor interactions.

mutant animals with corpse-clearance defects has led to the identification of eight components of the apoptotic-cell-body removal machinery, which are grouped with other genes that regulate apoptosis in the worm and are therefore termed *ced*, or *C. elegans* <u>d</u>eath, genes. Interestingly, most of the apoptotic-cell-clearance genes identified in *C. elegans* appear to encode signalling molecules. An exception to this may be Ced-1, which is localized to the plasma membrane and has homology with the scavenger receptor family [31], and which has an intracellular signalling domain with homology with CD91.

Despite the power of genetic manipulation in *C. elegans*, how each of the Ced-encoded molecules participates in the engulfment process will require further investigation. The *C. elegans* engulfment genes have been segregated into two groups based on complementation analysis. Perhaps surprisingly, neither single mutations nor any combination of mutations results in a complete abolishion of apoptotic-cell removal, reiterating the idea that the phagocytosis of apoptotic cells can be achieved through multiple redundant mechanisms. While the Ced-2 (CrkII)/Ced-5 (Dock180)/Ced-12 (ELMO; <u>e</u>ngulfment and cell <u>mo</u>tility)/Ced-10 (Rac) complementation group clearly represents components of the molecular machinery responsible for actin cytoskeleton rearrangement, how this pathway is activated and regulated is entirely unknown. However, over-expression of some of these molecules results in membrane ruffling, consistent with the concept that efferocytosis is macropinocytosis-like [32,33]. Curiously, many of the components of this second complementation group, as well as many of the receptors implicated in apoptotic-cell clearance, are also known to be involved in regulation of cholesterol homoeostasis. Whether these two groups in fact represent redundant signal-transduction pathways for apoptotic cell engulfment or whether they represent a single pathway with multiple points for regulation remains to be determined.

The complacent diner: implications of apoptotic-cell ingestion for the regulation of inflammation

Cytotoxic, pro-inflammatory and potentially auto-immunogenic debris can either be released from cells which have suffered an untimely necrotic death or leak from uningested apoptotic cells following secondary lysis. Therefore, phagocytosis of apoptotic cells while the cell body remains intact is important to prevent exposure of tissues to the noxious cytoplasmic contents from dying cells [3]. In addition, interaction of phagocytes with apoptotic cells results in the release of anti-inflammatory and immunosuppressive cytokines and prostanoids, including transforming growth factor-β and prostaglandins E_2 [1] and I_2 (C. Freire de Lima, personal communication), which further serve to promote maintenance of tissue homoeostasis during normal cell-death processes.

The cellular interactions that govern the inflammatory response in the face of cell death are not well understood. Many of the receptors that are known to be important for phagocytosis of apoptotic cells, such as Mer, the PSR and $\alpha_v\beta_3$ integrin/CD36, also are known to modulate cytokine production. Importantly, completion of engulfment is not necessary for regulation of the inflammatory response. Interaction of the phagocyte with apoptotic cells or PS alone is sufficient to promote the release of anti-inflammatory cytokines and chemokines. In fact, the presence of apoptotic cells in the absence of clearance may exacerbate disease processes, and has recently been shown to contribute to the pathogenesis of emphysema [34].

As both apoptotic cells and necrotic cells expose PS, the decision to commit to either an anti-inflammatory or pro-inflammatory response does not appear to occur from recognition of the dying cell *per se*, but is probably regulated by exposure to intracellular contents released from lysed necrotic cells, such as heat-shock proteins and the DNA transcription co-factor HMG-1 (high motility group-1) [35]. Interestingly, the process of apoptosis itself may serve to protect from exposure to these elements. For example, HMG-1 in apoptotic cells is bound to condensed DNA and therefore protected from accidental release. Therefore, there may actually be a hierarchy for induction of inflammation with necrotic cells, not previously apoptotic cells, being the most potent inducers of an inflammatory response, while apoptotic and subsequently lysed cells have an intermediate potential for inflammatory stimulation, and apoptotic unlysed cells are non-phlogistic in nature [36].

Conclusion: removal of apoptotic cell bodies is an ambitious undertaking

Disease pathogenesis results from uncontrolled cell proliferation or uncontrolled inflammation. Just as correct regulation of cell death is important for control of cell growth and differentiation, efficient removal of dying cells is critical for control of the immune response. Under normal circumstances, phagocytosis of apoptotic cells occurs so rapidly, that *in situ* apoptotic cells are rarely observed outside of other cell bodies. Therefore, when excess numbers of apoptotic cells are observed, it is important to consider a defect in apoptotic cell clearance as a factor in the progression of disease, as well as alterations in the rate of cell death.

In contrast to mechanisms of phagocytosis that have evolved to remove micro-organisms as a critical line of host defense against infection, and are thus designed to initiate beneficial inflammatory responses, efferocytosis is an evolutionarily conserved phagocytic mechanism designed to protect us from exposure to ourselves. To study how apoptotic cells are removed is an attempt to understand a wide variety of intricate cell–cell and protein-complex interactions, and the consequences of these interactions for regulation of inflammation and maintenance of normal tissue integrity and function.

Because this cellular undertaking occurs in nearly all multicellular organisms, understanding the process of apoptotic-cell removal is likely to contribute to our understanding of the origins of phagocytosis itself and regulation of the innate and adaptive immune responses. While many signals on the apoptotic-cell surface are likely to contribute to the recognition of apoptotic cells and their subsequent engulfment, the presence of PS has been shown to be important for stimulating both the uptake of apoptotic cells through a process similar to macropinocytosis. Importantly, PS has also been shown to regulate the production and release of anti-inflammatory cytokines. It is of no surprise, then, to learn that a variety of obligate intracellular pathogens, including *Leishmania*, and also some types of neoplastic cells, expose PS, and perhaps use apoptotic-cell mimicry to avoid immunodetection. Therefore, the study of apoptotic-cell clearance may ultimately lead to an understanding of pathogen invasion and evasion strategies, as well as to insights into the progression of certain types of cancer.

Summary

* *Recognition and removal of apoptotic cells is important for the prevention of exposure to the cytotoxic and potentially auto-immunogenic contents of dying cells.*

- *Interaction of phagocytes with apoptotic cells, but not necessarily engulfment per se, elicits the production of anti-inflammatory cytokines, and serves to both suppress inflammation and promote resolution of the inflammatory response.*

- *ACAMPs are important for recognition and phagocytosis of apoptotic cells. Known ACAMPs include the alterations in glycosylation patterns, changes in cell-surface protein conformation or expression, and exposure and/or oxidation of PS.*

- *Many different receptors have been proposed to be involved in apoptotic-cell recognition and clearance. These receptors may have redundant functions, or may act as part of a large, multi-component, apoptotic-cell-recognition synapse to promote corpse removal.*

- *The process of apoptotic cell removal is likely to constitute an evolutionarily conserved and unique phagocytic mechanism.*

References

1. Savill, J. & Fadok, V. (2000) Corpse clearance defines the meaning of cell death. *Nature (London)* **407**, 784–788

2. Savill, J., Dransfield, I., Gregory, C. & Haslett, C. (2002) A blast from the past: clearance of apoptotic cells regulates immune responses. *Nat. Rev. Immunol.* **2**, 965–975

3. Ren, Y. & Savill, J. (1998) Apoptosis: the importance of being eaten. *Cell Death Differ.* **5**, 563–568

4. Scott, R.S., McMahon, E.J., Pop, S.M., Reap, E.A., Caricchio, R., Cohen, P.L., Earp, H.S. & Matsushima, G.K. (2001) Phagocytosis and clearance of apoptotic cells is mediated by Mer. *Nature (London)* **411**, 207–211

5. Manfredi, A.A., Rovere, P., Galati, G., Heltai, S., Bozzolo, E., Soldini, L., Davoust, J., Balestrieri, G., Tincani, A. & Sabbadini, M.G. (1998) Apoptotic cell clearance in systemic lupus erythematosus. I. Opsonization by antiphospholipid antibodies. *Arthritis Rheum.* **41**, 205–214

6. Henson, P.M., Bratton, D.L. & Fadok, V.A. (2001) Apoptotic cell removal. *Curr. Biol.* **11**, R795–R805

7. Fadok, V.A. & Chimini, G. (2001) The phagocytosis of apoptotic cells. *Semin. Immunol.* **13**, 365–372

8. Gumienny, T.L. & Hengartner, M.O. (2001) How the worm removes corpses: the nematode C. *elegans* as a model system to study engulfment. *Cell Death Differ.* **8**, 564–568

9. Wiegand, U.K., Corbach, S., Prescott, A.R., Savill, J. & Spruce, B.A. (2001) The trigger to cell death determines the efficiency with which dying cells are cleared by neighbours. *Cell Death Differ.* **8**, 734–746

10. Williamson, P., van den Eijnde, S. & Schlegel, R.A. (2001) Phosphatidylserine exposure and phagocytosis of apoptotic cells. *Methods Cell Biol.* **66**, 339–364

11. Dini, L. (2000) Recognizing death: liver phagocytosis of apoptotic cells. *Eur. J. Histochem.* **44**, 217–227

12. Moffatt, O.D., Devitt, A., Bell, E.D., Simmons, D.L. & Gregory, C.D. (1999) Macrophage recognition of ICAM-3 on apoptotic leukocytes. *J. Immunol.* **162**, 6800–6810

13. Brown, S., Heinisch, I., Ross, E., Shaw, K., Buckley, C.D. & Savill, J. (2002) Apoptosis disables CD31-mediated cell detachment from phagocytes promoting binding and engulfment. *Nature (London)* **418**, 200–203

14. Pradhan, D., Krahling, S., Williamson, P. & Schlegel, R.A. (1997) Multiple systems for recognition of apoptotic lymphocytes by macrophages. *Mol. Biol. Cell* **8**, 767–778

15. Platt, N., da Silva, R.P. & Gordon, S. (1998) Recognizing death: the phagocytosis of apoptotic cells. *Trends Cell Biol.* **8**, 365–372

16. Devitt, A., Moffatt, O.D., Raykundalia, C., Capra, J.D., Simmons, D.L. & Gregory, C.D. (1998) Human CD14 mediates recognition and phagocytosis of apoptotic cells. *Nature (London)* **392**, 505–509

17. Fadok, V.A., Bratton, D.L., Rose, D.M., Pearson, A., Ezekewitz, R.A. & Henson, P.M. (2000) A receptor for phosphatidylserine-specific clearance of apoptotic cells. *Nature (London)* **405**, 85–90

18. Ishimoto, Y., Ohashi, K., Mizuno, K. & Nakano, T. (2000) Promotion of the uptake of PS liposomes and apoptotic cells by a product of growth arrest-specific gene, GAS6. *J. Biochem. (Tokyo)* **127**, 411–417

19. Savill, J., Hogg, N., Ren, Y. & Haslett, C. (1992) Thrombospondin cooperates with CD36 and the vitronectin receptor in macrophage recognition of neutrophils undergoing apoptosis. *J. Clin. Invest.* **90**, 1513–1522

20. Hanayama, R., Tanaka, M., Miwa, K., Shinohara, A., Iwamatsu, A. & Nagata, S. (2002) Identification of a factor that links apoptotic cells to phagocytes. *Nature (London)* **417**, 182–187

21. Anderson, H.A., Maylock, C.A., Williams, J.A., Paweletz, C.P., Shu, H. & Shacter, E. (2003) Serum-derived protein S binds to phosphatidylserine and stimulates the phagocytosis of apoptotic cells. *Nat. Immunol.* **4**, 87–91

22. Nauta, A.J., Daha, M.R., Kooten, C. & Roos, A. (2003) Recognition and clearance of apoptotic cells: a role for complement and pentraxins. *Trends Immunol.* **24**, 148–154

23. Ogden, C.A., deCathelineau, A., Hoffmann, P.R., Bratton, D., Ghebrehiwet, B., Fadok, V.A. & Henson, P.M. (2001) C1q and mannose binding lectin engagement of cell surface calreticulin and CD91 initiates macropinocytosis and uptake of apoptotic cells. *J. Exp. Med.* **194**, 781–795

24. Hoffmann, P.R., deCathelineau, A.M., Ogden, C.A., Leverrier, Y., Bratton, D.L., Daleke, D.L., Ridley, A.J., Fadok, V.A. & Henson, P.M. (2001) Phosphatidylserine (PS) induces PS receptor-mediated macropinocytosis and promotes clearance of apoptotic cells. *J. Cell Biol.* **155**, 649–659

25. Wood, W., Turmaine, M., Weber, R., Camp, V., Maki, R.A., McKercher, S.R. & Martin, P. (2000) Mesenchymal cells engulf and clear apoptotic footplate cells in macrophageless PU.1 null mouse embryos. *Development* **127**, 5245–5252

26. Aderem, A. & Underhill, D.M. (1999) Mechanisms of phagocytosis in macrophages. *Annu. Rev. Immunol.* **17**, 593–623

27. Swanson, S.J. & Watts, C. (1995) Macropinocytosis. *Trends Cell Biol.* **5**, 424–428

28. Leverrier, Y. & Ridley, A.J. (2001) Requirement for Rho GTPases and PI3-kinases during apoptotic cell phagocytosis by macrophages. *Curr. Biol.* **11**, 195–199

29. Grimmer, S., van Deurs, B. & Sandvig, K. (2002) Membrane ruffling and macropinocytosis in A431 cells require cholesterol. *J. Cell Sci.* **115**, 2953–2962

30. Albert, M.L., Kim, J.I. & Birge, R.B. (2000) Alpha$_v$beta$_5$ integrin recruits the CrkII-Dock180-Rac1 complex for phagocytosis of apoptotic cells. *Nat. Cell Biol.* **2**, 899–905

31. Zhou, Z., Hartwieg, E. & Horvitz, H.R. (2001) Ced-1 is a transmembrane receptor that mediates cell corpse engulfment in *C. elegans*. *Cell* **104**, 43–56

32. Tosello-Trampont, A.C., Brugnera, E. & Ravichandran, K.S. (2001) Evidence for a conserved role for CrkII and Rac in engulfment of apoptotic cells. *J. Biol. Chem.* **276**, 13797–13802

33. Brugnera, E., Haney, L., Grimsley, C., Lu, M., Walk, S.F., Tosello-Trampont, A.C., Macara, I.G., Madhani, H., Fink, G.R. & Ravichandran, K.S. (2002) Unconventional Rac-GEF activity is mediated through the Dock180-ELMO complex. *Nat. Cell Biol.* **4**, 574–582

34. Tuder, R.M., Petrache, I., Elios, J.A., Voelkel, N.T. & Henson, P.M. (2003) Apoptosis and emphysema: the missing link. *Am. J. Respir. Cell. Mol. Biol.* **28**, 555–562

35. Basu, S., Binder, R.J., Suto, R., Anderson, K.M. & Srivastava, P.K. (2000) Necrotic but not apoptotic cell death releases heat shock proteins, which deliver a partial maturation signal to dendritic cells and activate the NF-κB pathway. *Int. Immunol.* **12**, 1539–1546

36. Voll, R.E., Herrmann, M., Roth, E.A., Stach, C., Kalden, J.R. & Girkontaite, I. (1997) Immunosuppressive effects of apoptotic cells. *Nature (London)* **390**, 350–351

9

Apoptosis in disease: about shortage and excess

Thomas Brunner[1] and Christoph Mueller

Division of Immunopathology, Institute of Pathology, University of Bern, PO Box 62, Murtenstrasse 31, 3010 Bern, Switzerland

Abstract

The death of cells by apoptosis is a fundamental event in development and the maintenance of cell homoeostasis. The other side of the coin, however, is that excessive cell death by apoptosis or the lack of apoptosis is often the driving force of many diseases. Whereas reduced apoptosis sensitivity is a basic characteristic of many tumour cells, accelerated tissue cell death and loss of tissue functions is the underlying cause of many auto-immune and inflammatory diseases.

Introduction

In multicellular organisms the total number of cells has to be tightly regulated, both quantitatively and qualitatively. Various examples demonstrate that either uncontrolled cell growth or cell depletion may result in significant loss of vital functions, development of various diseases and possibly even the death of the affected individual. Whereas necrotic changes in tissues have been and still are considered an important aspect of various pathologies in humans, there is accumulating evidence that impaired cell death induced by apoptosis is crucially involved in many diseases. In this review, we will discuss the role of excessive and reduced apoptosis in the induction of disease in general, and will elaborate further this issue around T-cell-mediated immunopathologies.

[1] *To whom correspondence should be addressed*
(e-mail tbrunner@pathology.unibe.ch).

Apoptosis and disease: too little, too much...

Apoptosis inhibition is involved in tumour formation

Disease is a pathological disturbance of the normal physiological status, with severe consequences for the integrity of the affected organ or individual where the physiological processes are temporarily, or permanently, out of their dynamic balance. Thus, whenever apoptotic cell death contributes to the development of disease, this may be due to either too much, or too little, apoptosis. There are numerous examples for either situation in the various pathologies. If reduced apoptosis is the underlying cause of a given disease, we have to assume that the cell population inappropriately surviving may significantly harm the surrounding tissue and directly affect its function (Figure 1). A prime example for such a pathological situation is the formation of tumours. Tumour cells are genetically altered by mutations, which give rise to a cell population that is (relatively) insensitive to growth control and apoptosis induction. Although tumour cells are usually not directly cytotoxic

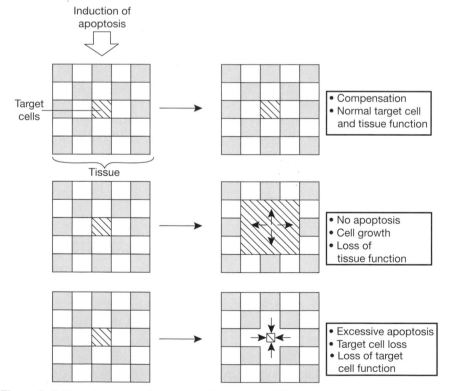

Figure 1. Reduced or excessive apoptosis can cause disease
If apoptosis induction can be compensated, normal tissue functions are preserved and disease does not develop. Reduced apoptosis sensitivity of a target-cell population, combined with uncontrolled growth, may cause damage to the neighbouring tissue. Excessive apoptosis may lead to loss of a target-cell population and its effector functions.

for normal parenchymal cells, they do compete for blood supply, growth factors and space, thus gradually affecting the viability and function of the neighbouring normal tissue. For example, a brain tumour may displace normal brain tissue, leading to severe loss of its function, but surgical removal of the tumour may often allow resumption of normal brain functions.

Why do tumour cells grow in an apparently uncontrolled manner? Obviously, the reduced control of cell-cycle progression and inappropriate growth factor expression contribute to this cell behaviour. Mutations in cell-cycle control or growth-factor-receptor genes leading to uncontrolled growth are often found in various tumours. However, there is increasing evidence that enhanced proliferation is compensated in normal cells by enhanced cell death, and that successful tumour formation frequently requires mutations that enhance the resistance to apoptosis induction. Evidence for this hypothesis can be tested in most tumour models, and is elegantly illustrated in a recent experimental study by Evan and co-workers [1]. In many tumours, the transcription factor and oncogene c-Myc is over-expressed and supports cell-cycle progression. Similarly, transgenic over-expression of c-Myc in pancreatic β-cells of the murine pancreas causes an initial burst of β-cell proliferation, which is, however, followed by extensive induction of apoptosis. Thus enhancement of cell growth is not sufficient for tumour formation, but may instead result in subsequent cell death. In contrast, if the anti-apoptotic molecule Bcl-x_L is co-expressed with c-Myc in β-cells, massive expansion of the transformed cells is observed with the formation of angiogenic, invasive tumours (insulinomas) [1]. Thus resistance to apoptosis induction is a key event in certain tumour development. Apoptosis-related genes are frequently mutated in tumour cells. Often, anti-apoptotic gene products are over-expressed, whereas pro-apoptotic genes are inactivated by mutations or deletions. It is thus not surprising that the first anti-apoptotic gene identified, Bcl-2, was isolated from a B-cell lymphoma [2]. Similarly, the tumour-suppressor-gene product p53 can induce apoptosis and p53 mutations are among the most common genetic alterations in malignant tumours [3].

Reduced T-cell apoptosis as an underlying defect in auto-immunity

Resistance to apoptosis is not only a central element in tumour formation, but is also involved in the pathogenesis a variety of other diseases. The ability of T-cells to die by apoptotic cell death is as important for induction of a proper immune response as is the induction of T-cell activation and proliferation. Recent years of research have substantiated the role of members of the TNF (tumour necrosis factor) family in the maintenance of immune homoeostasis and in the induction of T-cell apoptosis. In particular, Fas ligand (also called CD95 or APO-1 ligand) plays a predominant role in the regulation of T- and B-cell homoeostasis. Failure of mature T- and B-cells to undergo apoptosis is closely associated with development of auto-immunity [4]. Interestingly, mutations in the Fas receptor or Fas ligand are frequently found in a group of patients, suffering from the so-called ALPS (auto-immune lymphoproliferative syndrome). This disease is characterized by uncontrolled T- and B-cell proliferation, splenomegaly and formation of auto-antibodies, and strongly resembles the phenotype of the Fas-receptor-knockout mouse (reviewed in [5]). In many other auto-immune diseases, including rheumatoid arthritis [6], reduced T-cell apoptosis has also been suggested as one of the underlying defects leading to induction of disease.

Excessive target-cell apoptosis in disease

In contrast, in organ-specific auto-immune diseases, uncontrolled excessive target-cell apoptosis may be the disease-initiating event. As shown in Figure 1, excessive induction of apoptosis in a distinct target-cell population may gradually lead to their disappearance and loss of their effector functions. Type I diabetes (also termed juvenile diabetes or insulin-dependent diabetes mellitus) is a frequent auto-immune disease, in which self-reactive T-cells cause the destruction of the β-cells in the pancreas. As a result, the insulin-producing cells gradually disappear and affected patients have to compensate for reduced endogenous production by insulin injections. Why do autoreactive T-cells suddenly attack β-cells of the pancreas and cause their destruction? A likely possibility is that the central deletion process in the thymus, responsible for the elimination of the vast majority of autoreactive immature T-cells, is not as stringent as necessary. Thus, in most individuals, the presence of T-cells specific for β-cell antigens can be demonstrated in the peripheral blood. Overt type I diabetes, however, only develops in less than 1% of the Caucasian population, implicating that normally other mechanisms prevent the onset of the disease. Hengartner and colleagues [7] have demonstrated that viral antigens can be over-expressed in the pancreas of fully immunocompetent mice, without developing any pathological alterations. In contrast, upon challenge with the infectious virus, the immune system is stimulated and the resulting immune response is capable of eliminating the viral load, but the virus-specific T-cells now also attack the virus-protein-expressing β-cells, resulting in overt type I diabetes [7] (Figure 2). This experimental model system illustrates that self-tolerance can be broken by the induction of a potent pro-inflammatory immune response. Similar observations are made in other auto-immune diseases. For example, rheumatoid arthritis or EAE (experimental allergic encephalomyelitis), a disease similar to multiple sclerosis in humans, can be induced experimentally in rodents by injecting self-antigen in an inflammatory context, i.e. by co-injecting complete Freund's adjuvant. This induces an inflammatory response and causes the breakdown of self tolerance. Therefore, strong inflammatory responses and viral infections may often be initial triggers leading to auto-immune disorders.

Cell-mediated cytotoxicity has most probably not evolved to destroy tissue cells expressing self-antigen, but to defend the host against pathogens hiding within the cells of the body, such as viruses and intracellular bacteria.

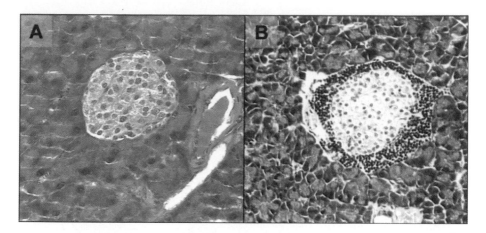

Figure 2. T-cell infiltration of the pancreatic β-cell islets during diabetes
NOD (non-obese diabetic) mice develop spontaneously auto-immune diabetes. (**A**) Normal pancreatic β-cell islet. (**B**) β-Cell islet with strong T-cell infiltrate in a diseased animal.

However, the same mechanisms responsible for the elimination of infected target cells also mediate the death of innocent tissue cells. T-cells kill their targets predominately via death receptor ligation and the perforin/granzyme B pathway (reviewed in [8]). Activated T-cells and natural killer cells express death ligands, like Fas ligand, TNFα and TRAIL (TNF-related apoptosis-inducing ligand). Upon receptor ligation on the corresponding target cell, the DISC (death-inducing signalling complex) forms and caspases (cysteine proteases with a preference for aspartic acid) become activated. Caspases are the major effector proteases in apoptosis induction, responsible for the demolition of the cell. Details of the signalling events that are initiated upon death receptor ligation have been extensively reviewed elsewhere [9] and will not be further discussed here. Although the initial events between death receptor-induced apoptosis and perforin/granzyme-B-induced target-cell killing are different, they both eventually activate the same suicide programme in the target cell. After the release of the cytotoxic granules from the activated T-cell, granzyme B binds to the target cells through the mannose-6-phosphate receptor and translocates with the help of perforin to the cytoplasm. Granzyme B, itself a protease, can cleave and activate caspases and initiate the apoptosis programme. Thus both cytotoxic effector mechanisms engage the target cell's own suicide programme (Figure 3) [8].

The question remains why different cytotoxic effector mechanisms are used to eventually engage the same signalling events, namely the apoptosis machinery. Target cells not only express different patterns and levels of death receptors, a prerequisite for the induction of apoptosis via this pathway, but also different patterns and levels of apoptosis inhibitors, capable of inhibiting either the death-receptor- or cytotoxic-granule-mediated pathway. For example, the experimental inhibition of the Fas pathway has a marked beneficial

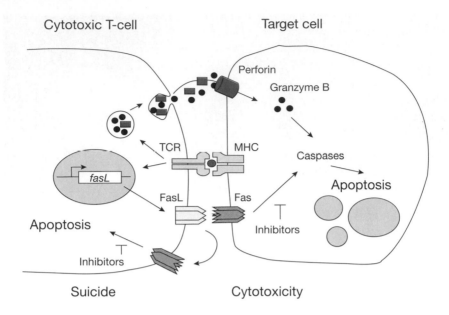

Figure 3. Mechanisms of T-cell-mediated target-cell apoptosis
T-cells recognize antigen presented on the MHC (major histocompability complex) through their specific TCR (T-cell receptor). T-cell activation leads to the release of cytotoxic granules, containing perforin and granzyme B. Perforin enables granzyme B to enter the target cell, and activate caspases and the cellular apoptosis programme. T-cell activation can also induce the expression of Fas ligand (FasL). FasL can bind to the Fas receptor on the target cell and induce apoptosis. Alternatively, T-cell-expressed FasL may bind to the Fas receptor on the same cell or other T-cells and induce homoeostatic apoptosis (suicide). Apoptosis in effector T-cells as well as target cells is regulated by apoptosis inhibitors.

effect on the course of EAE, suggesting that Fas plays a crucial role in demyelination and neuron destruction [10]. Similarly, Fas-induced apoptosis appears to be the predominant mechanism of hepatocyte killing during virus-induced hepatitis. Hepatocytes are known to express relatively high levels of Fas, which is even further up-regulated in an inflammatory environment. In addition, upon injection of an agonistic anti-Fas antibody, mice rapidly succumb due to massive liver cell apoptosis [11]. Thus hepatocyte killing via the Fas pathway appears to dominate. In contrast, in the elimination of other virus-infected cells or transplant rejection, the perforin/granzyme B pathway appears to be more important.

Excessive tissue cell apoptosis is not only observed during auto-immune diseases or, as will be discussed below, during excessive inflammatory responses, but is also a characteristic of other diseases. The loss of CD4+ T-cells is a hallmark of HIV infection in humans. As a result, the host becomes immunodeficient due to the lack of T-cell help in the induction of protective immune responses against opportunistic infections. Recently, it was recognized that this loss of CD4+ T-cells is also mediated by apoptosis. CD4+ T-cells are either directly killed by the virus upon infection, or become innocent bystander targets of activated cytotoxic T-cells (reviewed in [12]). Thus, similar to the disorders described above, apoptosis induction in a distinct target-cell population (CD4[+] T-cells) may result in severe consequences for the HIV-infected individual.

The examples discussed above indicate that either reduced or excessive cellular apoptosis can result in a severe impairment of specific tissue functions, often leading to disease (Figure 4). Whereas some tissues may be extremely sensitive to cell loss, others may well have a certain buffering capacity, i.e. can tolerate the loss of a significant number of cells due to subsequent compensation or regeneration. Disease may thus represent a situation where cell loss cannot be compensated anymore.

The intestinal mucosa: a site of death and life

The intestinal mucosa is constantly confronted with two apparently conflicting duties, i.e. to allow uptake of nutrients and to prevent invasion of the intestinal mucosa by the luminal microflora. This is certainly not a trivial task given the huge number (>10[14]) of bacteria present in the intestine, which even exceeds the total number of cells in the rest of the human body, and the total surface of the intestinal mucosa (approx. 300 m^2) that needs to be protected from invasive microbes. Furthermore, the luminal content may contain potentially toxic or mutagenic metabolites derived from ingested food and possibly also bacterially derived metabolites, which may threaten the integrity of the intestinal mucosa, especially the epithelium (reviewed in [13]). This makes it clear that local immune responses against invading pathogens have to be tightly regulated and need to be carried out with the appropriate immune mechanisms that limit the spread of infiltrating microbes, however, without causing excessive bystander damage to non-infected cells in the intestinal epithelium and lamina propria.

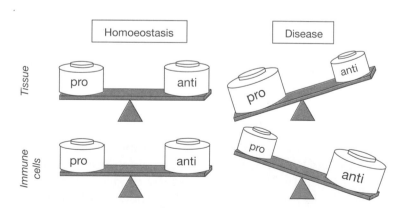

Figure 4. Disease as an unbalance between pro- and anti-apoptotic signals
Increased resistance to apoptosis induction in immune cells and decreased apoptosis resistance in tissue cells can cause disease.

The importance of a tightly regulated local immune system in the maintenance of tissue homoeostasis has been demonstrated clearly in mouse strains that are deficient for functional genes normally involved in mounting (and regulating) immune responses, e.g. IL-2 (interleukin 2), IL-10 and transforming growth factor β (reviewed in [14]). Intriguingly, in the presence of conventional bacterial flora, these mouse strains spontaneously develop a chronic inflammatory disorder of the intestine, primarily of the colon where the largest bacterial load in the intestinal lumen is found. Whereas the absence of the immunosuppressive cytokines IL-10 or transforming growth factor β may lead to an impaired development of regulatory T-cells (T-cells capable of controlling immune responses, e.g. via the production of immunosuppressive cytokines), and hence to an excessive expansion of inflammatory T-cells in the colonic mucosa, it is believed that impaired apoptosis induction in T-cells in the absence of IL-2 production is the underlying cause of colitis (inflammatory disease in the colon) induction in IL-2-deficient mice. IL-2 is well known as a T-cell growth factor. Surprisingly, T-cells in the IL-2-deficient mouse show no defects in proliferation, but reduced sensitivity to apoptosis. Chronically stimulated T-cells become increasingly sensitive to Fas receptor ligation. IL-2 not only enhances the expression of Fas ligand on activated T-cells and thus contributes to their elimination, but also down-regulates the specific inhibitor of the Fas signalling pathway cFLIP (cellular Flice-like inhibitory protein) [15]. cFLIP is a caspase-8 homologue that lacks its catalytic domain. It acts as a dominant-negative inhibitor and blocks the Fas signalling pathway at the receptor-complex level. Failure to produce sufficiently high levels of IL-2 and down-regulate cFLIP may thus lead to enhanced apoptosis resistance in effector T-cells, and eventually to uncontrolled immune reactions and severe tissue damage. Interestingly, impaired apoptosis induction in colitis-inducing T-cells is not only observed in experimental animal models, but also in human patients. Crohn's disease and ulcerative colitis are two main forms of inflammatory bowel disease in humans. Intestinal T-cells isolated from Crohn's disease and ulcerative colitis patients show reduced *ex vivo* Fas-sensitivity when compared with cells isolated from controls. Similarly, reduced *in situ* T-cell apoptosis is observed in the colonic mucosa of Crohn's disease patients (reviewed in [16]). These observations from human patients and experimental model systems strongly support the idea that T-cell apoptosis in the intestinal mucosa is an important immunomodulatory mechanism and has to be tightly regulated to avoid uncontrolled, destructive immune responses.

Altered apoptosis sensitivity may lead to inappropriate survival of intestinal cytotoxic T-cells and apoptosis induction in innocent bystander targets, e.g. epithelial cells. Epithelial cells are Fas-sensitive and are readily killed by Fas ligand-expressing T-cells. Apart from Fas ligand, another death-inducing ligand appears to play an important role in the pathogenesis of inflammatory bowel disease. TNFα is produced by activated T-cells and macrophages and can induce apoptosis upon binding to TNF receptors I and II and engaging the caspase cascade. Experimental work in animal models clearly demonstrated a crucial role for TNFα in the induction of inflammatory disorders of the intestine, such as colitis and graft-versus-host disease. The mechanism of TNFα-mediated disease induction has not yet been fully defined. However, the local release of TNFα leads to rapid apoptosis induction in epithelial cells, immediately followed by shedding of the dying cells into the intestinal lumen [17]. This strongly suggests that TNFα-induced epithelial cell apoptosis also affects the permeability of the intestinal epithelium. The uncontrolled entry of immunostimulatory bacterial products will activate macrophages and T-cells further, leading to a vicious circle of cell activation and tissue destruction. The central role for TNFα in the pathogenesis of inflammatory bowel disease in human patients, in particular in patients with Crohn's disease, has been demonstrated by the observed dramatic response rate in patients with active disease following administration of a humanized anti-TNFα antibody (Infliximab) [18]. Anti-TNFα probably produces its effect through the neutralization of TNFα; however, it is also possible that the TNFα-binding antibody mediates the specific depletion of cell-surface TNFα-expressing immune effector cells, e.g. via complement activation or natural-killer-cell activation through Fc receptors. Evidence for an apoptosis-inducing effect of Infliximab in transmembrane TNFα-expressing monocytes *in vitro* has been reported recently [19].

Apart from direct apoptosis induction in epithelial cells, TNFα may also affect the development of inflammatory bowel disease through other mechanisms. As mentioned above, epithelial cells express Fas and can be killed by Fas-ligand-expressing T-cells. Inflammatory cytokines, such as TNFα and interferon γ, not only enhance the expression of the Fas receptor, but also sensitize the cells for Fas-induced apoptosis, presumably by down-regulating anti-apoptotic molecules. Such sensitized epithelial cells may then become easy prey of activated cytotoxic T-cells, infiltrating the lamina propria and the epithelial cell layer during inflammatory bowel disease.

In summary, there is increasing evidence that apoptosis induction plays a crucial role in the development of inflammatory disease of the intestinal mucosa, such as Crohn's disease and ulcerative colitis. The inflammatory environment may lead to reduced homoeostatic T-cell apoptosis as well as enhanced target cell killing and tissue destruction (Figure 5).

Therapeutic approaches

Obviously, progress in the understanding of the role of apoptosis in the development of various diseases must also be translated into therapeutic approaches. Multiple approaches aim at preventing apoptosis of vital target cells during diseases characterized by rapid and excessive target cell death. For example, small-molecule caspase inhibitors may have promising protective effects on neuronal cell loss during stroke [20].

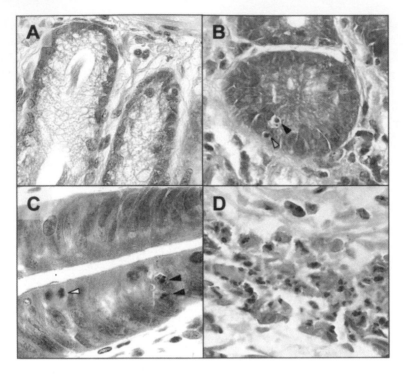

Figure 5. Uncontrolled T-cell activation in the intestinal mucosa causes excessive tissue cell apoptosis
(**A**) Tissue section through normal intestinal mucosa (mouse colon) showing the crypt section. (**B**) Detection of apoptotic nuclei (black arrowhead) in the epithelial layer during experimental colitis in the mouse. Note the epithelial cell layer infiltrating lymphocyte (grey arrowhead). (**C**) Compensation of excessive epithelial cell apoptosis (black arrowheads) by increased proliferation (cells in mitosis, white arrowhead). Experimental colitis in the mouse: (**D**) apoptotic epithelial cells are shed into the gut lumen during experimental colitis in the mouse.

In diseases characterized by excessive tissue cell apoptosis due to uncontrolled activation of T-cells, however, another approach may prove to be successful. The examples described above have shown that T-cell apoptosis may represent an interesting target. Compounds or neutralizing antibodies that enhance the apoptosis sensitivity of T-cells due to down-regulation of anti-apoptotic molecules, such as cFLIP or Bcl-x_L, have already been used successfully to inhibit disease induction in experimental animal models. For example, the protein kinase C inhibitor bisindolylmaleimide VIII can specifically down-regulate cFLIP expression, and *in vivo* administration has a strong ameliorating effect during experimental rheumatoid arthritis and EAE [4,21]. These experiments show that disease treatment by sensitizing T-cells to apoptosis induction is an interesting and applicable concept. However, a profound understanding of effector-cell apoptosis regulation may be required to specifically interfere with the pathogenesis of the various diseases.

Conclusion

A more profound understanding of the molecules and pathways leading to apoptotic cell death, and their regulation during the pathogenesis of different diseases, may allow the identification of crucial target molecules and promote the development of therapeutic protocols and compounds to specifically target a disease-inducing cell population or prevent target cell death.

Summary

- *In multicellular organisms, the balance between cell growth and cell loss has to be tightly regulated.*
- *Reduced or excessive cell death by apoptosis can lead to an imbalance of this cellular homoeostasis, and may lead to a loss of tissue function and eventually to disease.*
- *Uncontrolled T-cell activation can lead to excessive tissue cell apoptosis and loss of vital functions.*
- *Failure to control T-cell effector functions by apoptosis is an underlying cause of many diseases.*

We thank the members of our laboratories and the Swiss National Science Foundation, Oncosuisse, and the Crohn's and Colitis Foundation of America for ongoing support.

References

1. Pelengaris, S., Kahn, M. & Evan, G. (2002) Suppression of Myc-induced apoptosis in beta cells exposes multiple oncogenic properties of Myc and triggers carcinogenic progression. Cell **109**, 321–334
2. Pegoraro, L., Palumbo, A., Erikson, J., Falda, M., Giovanazzo, B., Emanuel, B., Rovera, G., Nowell, P. & Croce, C. (1984) A 14;18 and an 8;14 chromosome translocation in a cell line derived from an acute B-cell leukemia. Proc. Natl. Acad. Sci. U.S.A. **81**, 7166–7170
3. Hollstein, M., Shomer, B., Greenblatt, M., Soussi, T., Hovig, E., Montesano, R. & Harris, C. (1996) Somatic point mutations in the p53 gene of human tumors and cell lines: updated compilation. Nucleic Acids Res. **24**, 141–146
4. Brunner, T. & Mueller, C. (1999) Is autoimmunity coming to a Fas(t) end? Nat. Med. **5**, 19–20
5. Lenardo, M., Chan, K.M., Hornung, F., McFarland, H., Siegel, R., Wang, J. & Zheng, L. (1999) Mature T lymphocyte apoptosis–immune regulation in a dynamic and unpredictable antigenic environment. Annu. Rev. Immunol. **17**, 221–253
6. Pope, R. (2002) Apoptosis as a therapeutic tool in rheumatoid arthritis. Nat. Rev. Immunol. **2**, 527–535
7. Ohashi, P., Oehen, S., Buerki, K., Pircher, H., Ohashi, C., Odermatt, B., Malissen, B., Zinkernagel, R. & Hengartner, H. (1991) Ablation of "tolerance" and induction of diabetes by virus infection in viral antigen transgenic mice. Cell **65**, 305–317
8. Barry, M. & Bleackley, R.C. (2002) Cytotoxic T lymphocytes: all roads lead to death. Nat. Rev. Immunol. **2**, 401–409
9. Peter, M.E. & Krammer, P.H. (1998) Mechanisms of CD95 (APO-1/Fas)-mediated apoptosis. Curr. Opin. Immunol. **10**, 545–551

10. Sabelko-Downes, K.A., Cross, A.H. & Russell, J.H. (1999) Dual role for Fas ligand in the initiation of and recovery from experimental allergic encephalomyelitis. J. Exp. Med. **189**, 1195–1205

11. Ogasawara, J., Watanabe-Fukunaga, R., Adachi, M., Matsuzawa, A., Kasugai, T., Kitamura, Y., Itoh, N., Suda, T. & Nagata, S. (1993) Lethal effect of the anti-Fas antibody in mice. Nature (London) **364**, 806–809

12. Selliah, N. & Finkel, T. (2001) Biochemical mechanisms of HIV induced T cell apoptosis. Cell Death Differ. **8**, 127–136

13. Groux, H. & Powrie, F. (1999) Regulatory T cells and inflammatory bowel disease. Immunol. Today **20**, 442–445

14. Elson, C., Cong, Y., Brandwein, S., Weaver, C., McCabe, R., Mahler, M., Sundberg, J. & Leiter, E. (1998) Experimental models to study molecular mechanisms underlying intestinal inflammation. Ann. N.Y. Acad. Sci. **859**, 85–95

15. Refaeli, Y., Van Parijs, L., London, C.A., Tschopp, J. & Abbas, A.K. (1998) Biochemical mechanisms of IL-2-regulated Fas-mediated T cell apoptosis. Immunity **8**, 615–623

16. Neurath, M.F., Finotto, S., Fuss, I., Boirivant, M., Galle, P.R. & Strober, W. (2001) Regulation of T-cell apoptosis in inflammatory bowel disease: to die or not to die, that is the mucosal question. Trends Immunol. **22**, 21–26

17. Piguet, P.F., Vesin, C., Donati, Y. & Barazzone, C. (1999) TNF-induced enterocyte apoptosis and detachment in mice: induction of caspases and prevention by a caspase inhibitor, ZVAD-fmk. Lab. Invest. **79**, 495–500

18. Papadakis, K.A. & Targan, S.R. (2000) Tumor necrosis factor: biology and therapeutic inhibitors. Gastroenterology **119**, 1148–1157

19. Lugering, A., Schmidt, M., Lugering, N., Pauels, H.G., Domschke, W. & Kucharzik, T. (2001) Infliximab induces apoptosis in monocytes from patients with chronic active Crohn's disease by using a caspase-dependent pathway. Gastroenterology **121**, 1145–1157

20. Hara, H., Friedlander, R.M., Gagliardini, V., Ayata, C., Fink, K., Huang, Z., Shimizu-Sasamata, M., Yuan, J. & Moskowitz, M.A. (1997) Inhibition of interleukin 1beta converting enzyme family proteases reduces ischemic and excitotoxic neuronal damage. Proc. Natl. Acad. Sci. U.S.A. **94**, 2007–2012

21. Zhou, T., Song, L., Yang, P., Wang, Z., Lui, D. & Jope, R.S. (1999) Bisindolylmaleimide VIII facilitates Fas-mediated apoptosis and inhibits T cell-mediated autoimmune diseases. Nat. Med. **5**, 42–48

10

Therapeutic approaches to the modulation of apoptosis

Finbarr J. Murphy[1], Liam T. Seery and Ian Hayes

EiRx Therapeutics Ltd, 2800 Cork Airport Business Park, Kinsale Road, Cork, Ireland

Abstract

The appreciation of the role of apoptosis in the vast majority of diseases affecting humans has revolutionized the discovery and development of drugs targeting inflammation and oncology. Novel therapeutic approaches to modulate disease by regulating apoptosis are currently being tested in pre-clinical and clinical settings. Enthusiasm for some of these therapies is reflected in the fact that they have received U.S. Food and Drug Administration approval in record time. Approaches include the traditional use of small molecules to target specific players in the apoptosis cascade. They also include radical new approaches such as using antisense molecules to inhibit production of the Bcl-2 protein or antibodies that ligate either death receptors, such as TRAIL (tumour necrosis factor-related apoptosis-inducing ligand), or the MHC (HLA-DR), resulting in the initiation of apoptosis of target cells. Antibodies targeting cell-specific antigens are being used in conjunction with radioactive isotopes to deliver a more specific chemotherapy, particularly in the case of B-cell lymphomas. Other therapies target mitochondria, a key organelle in the apoptosis cascade. This diverse range of therapies includes photodynamic therapy, retinoic acid and arsenic trioxide, all of which induce apoptosis by generating reactive oxygen species. As our understanding of

[1]To whom correspondence should be addressed
(e-mail fmurphy@eirxtherapeutics.com).

apoptosis increases, further opportunities will arise for tailor-made therapies that will result in improved clinical outcome without the devastating side effects of current interventions.

Introduction

Targeting apoptotic processes for therapy is novel only in terms of our understanding that this process underlies many cellular drug responses. Chemotherapeutic and radiation regimes, which have been the traditional mainstays of cancer therapy for several decades, have generally been regarded as successful if they have controlled tumour growth by killing transformed cells. However, it is only over the last decade that scientists and clinicians have realized that the desired effect of tumour shrinkage occurred through tumour cell apoptosis. In a similar fashion, corticosteroids that have been used as anti-inflammatory agents for many years have only very recently had their mode of action realized. Again, these effects are mediated largely by translocation of cytoplasmic steroid receptors to the nucleus that then signal apoptosis.

The consequences of therapeutic modulation of apoptosis will have profound implications in medical treatment for a number of diseases where there is either too little or too much cell death (Table 1). Understanding the pathways involved offers the possibility of manipulating cell systems in ways never before attempted. This will undoubtedly result in a wave of novel therapeutics, some of which are already in clinical trials. It is anticipated that as the processes of apoptosis are elucidated, novel cell- and context-specific regulators will be identified. These candidate therapeutic targets will fuel the impetus for innovative routes to apoptotic therapy. While this review will examine advances made with current strategies, it will also extend into areas where some of the more exciting approaches to therapeutics are beginning to emerge, as summarized in Table 2.

Table 1. Diseases caused by aberrant apoptosis

Too much apoptosis	Too little apoptosis
Alzheimer's disease	Cancer
Parkinson's disease	Atopic dermatitis
Huntington's disease	Asthma
Multiple sclerosis	Crohn's disease
Insulin-dependent diabetes mellitus	Rheumatoid arthritis
Hashimoto's thyroiditis	Chronic obstructive pulmonary disease (COPD)
Myocardial infarction	Osteoporosis
Stroke	
Brain/spinal injury	
Sepsis	
Graft versus host disease	

Table 2. Summary of novel therapeutic approaches to modify apoptosis

NF-κB, nuclear factor κB; PI 3-kinase, phosphoinositide 3-kinase.

Mechanism of action	Company	Product name	Indication
Caspase inhibitors	Idun Pharmaceuticals	IDN-6556	Hepatic disease
		IDN-6734	Myocardial infarction
	Vertex Pharmaceuticals	Vx-799	Sepsis
Histone deactylase inhibitor	Titan Pharmaceuticals	Pivanex	Non-small-cell lung carcinoma
Proteasome inhibitors	Millennium Pharmaceuticals	Velcade (Bortezomib, formerly LDP-341, PS-341)	Multiple myeloma
NF-kB inhibitor	Aventis Pharma	PS-1145	Multiple myeloma
PI 3-kinase/Akt inhibitors	ProlX Pharmaceuticals	DPI analogues	Cancer
	Kyowa Hakko Kogyo Co.	UCN-01 (7-hydroxystaurosporine)	Cancer
Farnesyl transferase inhibitors	Ortho Biotech Oncology	Zarnestra (formerly R115777)	Multiple myeloma
	Schering-Plough	Sarasar (formerly SCH66336)	Leukaemia, lung cancer
Tyrosine kinase inhibitors	Novartis Pharmaceuticals	Gleevac (Imatinib Mesylate)	Chronic myelogenous leukaemia
	Novartis Pharmaceuticals	PKI166	Oral cancer
Death receptors	Amgen/Genentech	TRAIL (Apc2)	Glioma

contd ☞

Table 2. (contd)

Mechanism of action	Company	Product name	Indication
Antibodies	Genentech/IDEC	Rituximab	Non-Hodgkin's lymphoma
	Corixa/GlaxoSmithKline	Bexxar (131I Tositumomab)	Non-Hodgkin's lymphoma
	IDEC/Schering	Zevalin	Non-Hodgkin's lymphoma
	Protein Design Labs/Exelixis	Remitogen	Non-Hodgkin's lymphoma
	Genentech	Herceptin (Trastuzumab)	Breast cancer
Antisense	Genta/Aventis Pharma	Genasense (G-3139)	Chronic lymphocytic leukaemia, non-small-cell lung carcinoma, malignant melanoma
Gene therapy	Canji/Schering-Plough	SCH 58500	Ovarian cancer
	Introgen Therapeutics	Advexin (formerly INGN 201)	Head and neck cancer
	Onyx Pharmaceuticals	ONYX-015	Cancer
Photodynamic therapy	QTL PhotoTherapeutics	Photofrin	Oesophageal cancer, microinvasive lung cancer
Retinoic acid	Johnson & Johnson	Fenretinide (4HPR)	Breast cancer
	Galderma Laboratories	AHPN (CD437)	Lung cancer
Arsenic trioxide	Cell Therapeutics	Trisenox	Acute promyelocytic leukaemia

Small molecules

Recent advances in the key arenas of high-throughput cell-based screening and synthesis of combinatorial libraries, complemented by *in silico* analysis, have facilitated the development of classical small-molecule drugs that can regulate the apoptosis cascade.

Caspase inhibitors

One of the first biochemical and detected consequences of cells undergoing apoptosis is the activation of cysteinyl aspartate-specific proteases, termed caspases, of which there are 14 known members to date (Table 3). Caspase activity results in the proteolytic cleavage of many key proteins involved in cellular metabolism. Consequently, caspases were one of the first targets explored for therapeutic modulation of apoptosis. Initial studies using active-site-mimetic peptide ketones, such as benzyloxycarbonyl-VAD-fluoromethylketone (z-VAD-fmk), benzyloxycarbonyl-YVAD-fluoromethylketone/-chloromethylketone (z-YVAD-fmk/z-YVAD-cmk) and *N*-benzyloxycarbonyl-DEVD-fluoromethylketone (z-DEVD-fmk) provided proof of principle that caspase inhibition improved survival, decreased infarct volume and improved organ function in animal models of ischaemia and myocardial infarct [1–3]. These studies suggested that caspase inhibition should be therapeutically effective for treatment of several diseases.

A number of large pharmaceutical and biotechnology companies have entered the race to identify various membrane permeable caspase-specific and pan-caspase inhibitors for a number of indications. Idun Pharmaceuticals, for example, have a number of caspase inhibitors in clinical trials, including IDN-

Table 3. Caspase family members

ICE, interleukin-1β-converting enzyme.

Caspase	Description
1	Involved in processing of interleukins/ICE subfamily
2	Mediator of cell-stress-induced apoptosis
3	Terminal executioner caspase
4	Inflammation/apoptosis
5	Inflammation/apoptosis
6	Terminal executioner caspase
7	Terminal executioner caspase
8	Mediates death-receptor apoptosis
9	Mediates mitochondrion-specific apoptosis
10	Mediates death-receptor apoptosis
11	Mediator of septic shock in mice
12	Mediates endoplasmic-reticulum-specific apoptosis
13	Inflammation/apoptosis
14	Expressed during epidermal differentiation

6556 for hepatic disease (phase II) and IDN-6734 for acute myocardial infarction (phase I). Another indication where caspase inhibitors may prove extremely beneficial is sepsis. Sepsis is a severe life-threatening bacterial infection of the bloodstream that overwhelms the body's immune system and remains one of the major killers in intensive-care units, with mortality rates changing little in the last decade. One characteristic of sepsis is the huge amount of apoptosis seen in lymphocytic cells. Recently, it was shown that inhibition of caspases prevented lymphocyte apoptosis and improved survival in mice. From these studies, it was concluded that acute treatment with caspase inhibitors acts to control infections by preventing lymphocyte apoptosis [4] (Figure 1). Presently, Vertex Pharmaceuticals is moving their small-molecule caspase inhibitor Vx-799 into pre-clinical trials as a potential therapy for sepsis.

One caveat that might limit the use of caspase inhibitors pertains to the extent of caspase-dependent apoptosis. Although initially thought to be involved in all pathways of apoptosis, more recently, it has been realized that caspases may have a more limited role. This is particularly evident in models of brain ischaemia, where there seems to be a continuum of death that ranges from caspase-dependent to caspase-independent pathways. Indeed, it has been suggested that that the developing brain moves away from caspase-mediated cell death to ensure tighter controls on events that might prove catastrophic if the system goes awry [5]. In support of this hypothesis are the findings of other investigators that several of the caspases that are detected in the embryo are not detected in the adult animal [6–8].

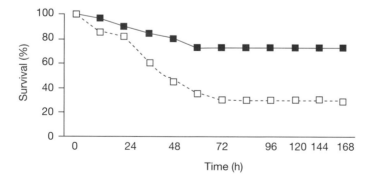

Figure 1. Protective effect of caspase inhibitor on survival
ND4 mice underwent caecal ligation and received antibiotics. After 90 min, mice received 10 mg/kg pan-caspase inhibitor (■) or diluent (□) and were studied over the indicated period. There was a significant decrease in the number of mice dying from sepsis in the treated group. Reproduced from Hotchkiss, R.S., Chang, K.C., Swanson, P.E., Tinsley, K.W., Hui, J.J., Klender, P., Xanthoudakis, S., Roy, S., Black, C., Grimm, E. et al. (2000) Caspase inhibitors improve survival in sepsis: a critical role of the lymphocyte. *Nat. Immunol.* **1**, 496–501, with permission from the Nature Publishing Group (www.nature.com/ni).

Bcl-2

Bcl-2 and related family members (including death antagonists, such as Bcl-2 and Bcl-x_L, and the death agonists Bax, Bak, Bid and Bad) are critical regulators of apoptosis. A high level of Bcl-2 protein is found in a large number of human cancers [9]. Regulation of apoptosis by protein–protein interactions between Bcl-2 family members is mediated via a hydrophobic binding pocket. By using a computer screening strategy based on the predicted structure of Bcl-2, Wang et al. [10] identified a small non-peptide ligand termed HA14-1 thought to bind the BH3 (Bcl-2 homology 3) domain of Bcl-2 and inhibit dimerization. Human acute myeloid leukaemia (HL-60) cells over-expressing Bcl-2 protein were observed to undergo apoptosis (>90%) at a dose of 50 µM HA14-1.

Targeting protein–protein interactions with small molecules has historically proven to be very difficult. Indeed, more recent studies on HA14-1 suggest that, although it opens the mitochondrial permeability transition pore, leading to mitochondrial depolarization and apoptosis, its mode of action is independent of BH3 binding. Instead it is thought to disrupt ADP/ATP fluxes in a redox-dependent manner that results in mitochondrial depolarization (F.E. Cotter, personal communication). Additionally, current studies *in vivo* demonstrate that HA14-1 is not well tolerated, suggesting that it will not be suitable for studies in humans (F.E. Cotter, personal communication). Studies are ongoing to uncover small-molecule variants of HA14-1 that will be more specific and less toxic.

p53

To identify a compound that could block the action of p53, Komarov et al. [11] used a reporter gene that turns cells blue when p53 is activated. From a screen of 10 000 compounds, the team identified a number of compounds that suppressed p53, even though the cells were exposed to the p53-inducing chemotherapeutic agent doxorubicin. One compound was chosen, named pifithrin α (an abbreviation for 'p53 inhibitor'), which did not affect cell growth or survival. Intraperitoneal injection of pifithrin α (2.2 mg/kg) was able to protect mice from a radiation dose that would normally kill 60% of the animals. Some mice survived even higher doses of radiation when given pifithrin α (Figure 2), and weight loss in mice treated with pifithrin α was also compared with the few remaining irradiated survivors that did not receive the drug. Administration of pifithrin α protected mice from the lethal genotoxic stress associated with anti-cancer treatment without promoting the formation of tumours, suggesting that inhibition of p53 maybe used to counteract the side effects of chemotherapy [11]. More recently, pifithrin α and a modified version of it called Z-1-117 have caused much excitement by their ability to protect midbrain dopaminergic neurons and improving behavioural outcome in a mouse model of Parkinson's disease [12].

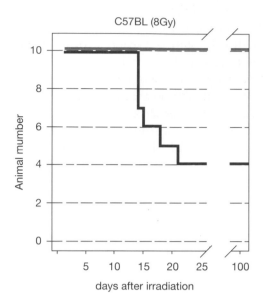

Figure 2. Survival conferred by administration of pifithrin α against whole-body γ irradiation in C57BL6 mice

Mice received a single intraperitoneal injection of 2.2 mg of pifithrin α/kg of body mass. The blue line indicates survival of treated mice relative to that of untreated mice (black line). Reprinted with permission, from Komarov, P.G., Komarova, E.A., Kondratov, R.V., Christov-Tselkov, K., Coon, J.S., Chernov, M.V. & Gudkov, A.V. (1999) A chemical inhibitor of p53 that protects mice from the side effects of cancer therapy. *Science* **285**, 1733–1737. Copyright (1999) American Association for the Advancement of Science.

Modulators of gene expression

HDAC (histone deactylase)-directed therapy

Histone acetylation, mediated by a family of proteins known as the HDACs, is critical for the proper transcriptional regulation of genetic programmes underlying virtually every major cell function including cell growth, differentiation, apoptosis, DNA repair and cell–cell/cell–substratum interactions. In general, chromatin fractions enriched in actively transcribed genes are also enriched in highly acetylated core histones, while under-acetylated histones are characteristic of transcriptionally inactive regions.

HDAC inhibitors cause acetylated histones to accumulate in both tumour and normal tissues. Depending on the cell type, this build up of acetylated histones in cancer cells can lead to transcriptional activation, cell-cycle arrest, differentiation or induction of apoptosis. Since subsets of HDACs bind to and regulate discrete subsets of genes, it is likely that the identification of selective HDAC inhibitors will provide better therapies for human cancer. However, as the number of HDACs is limited, this approach will not allow targeting of single genes. Also, since HDAC inhibitors cause acetylated histones to accumulate in both tumour and normal tissues, bystander effects in healthy tissue may give rise to unacceptable toxicity.

The HDAC inhibitor Pivanex, an analogue of butyric acid, is currently undergoing phase III clinical trials for non-small-cell lung carcinoma. In a multicentre, open-label study, scientists administered Pivanex to 47 patients who had advanced non-small-cell lung carcinoma and who had failed prior chemotherapy. Pivanex treatment resulted in a 1-year survival rate of 47% and a median survival rate of 11.1 months in 29 patients whose cancer had progressed after one or two prior chemotherapy regimens, as compared with the 1-year survival rate of 37% and a median survival rate of 7.5 months historically seen with use of docetaxel (Taxotere). Patients treated with Pivanex in this preliminary phase II study showed decreased pleural effusions, weight gain, decreased coughing and resolution of blood expectoration (haemoptysis) [13]. In addition, Pivanex has shown potential as a selective agent for haematopoietic malignancies including those that have acquired resistance to conventional chemotherapies [14].

Proteasome inhibitors

The 26 S proteasome, an ATP-dependent multicatalytic protease, is responsible for most of the ubiquitination-mediated degradation of intracellular proteins in eukaryotic cells. As well as removing damaged or misfolded proteins, the ubiquitin/proteasome pathway plays an important regulatory role in several key functions of the cell, including the cell cycle, neoplastic growth, metastasis and apoptosis. One reason why this pathway is so critical to the cell is its role in regulating transcription. The ubiquitin/proteasome pathway degrades IκB (inhibitory κB), allowing NF-κB (nuclear factor κB) to move into the nucleus and transcribe key cell-survival genes. The small-molecule proteasome inhibitor Velcade blocks the degradation of IκB, resulting in inhibition of NF-κB activation. *In vitro* results confirm that Velcade induces apoptosis through caspase-8 and caspase-9 activation in drug-resistant multiple myeloma cell lines and patients' cells [15]. *In vivo*, Velcade inhibits the growth of human multiple myeloma cells, decreases tumour-associated angiogenesis and prolongs the survival of SCID mice bearing human multiple myeloma cells [16]. Additionally, a combination of Velcade and the heat-shock protein 90 inhibitor geldanamycin has been reported to enhance the Velcade-triggered multiple myeloma-cell apoptosis [17].

In May 2003, after one of the most rapid approval procedures for a cancer drug, Velcade was approved by the FDA (U.S. Food and Drug Administration) for the treatment of myeloma patients who have received at least two prior therapies and have demonstrated disease progression on the last therapy. Velcade is also under investigation in phase I and II studies for a wide variety of solid cancers, including prostate, colorectal and lung cancer.

NF-κB inhibition

Preventing NF-κB activity by directly inhibiting IkB kinase (and thus preventing IκB degradation) is also being examined as a strategy for apoptosis modulation. One such inhibitor, PS-1145, has shown activity in an *in vitro* study of multiple myeloma. Studies have shown that PS-1145 prevented TNFα (tumour necrosis factor α)-induced NF-κB activation in a dose- and time-dependent fashion through inhibition of phosphorylation and subsequent degradation of IκBα. While PS-1145 only inhibited multiple myeloma cell proliferation by 20–50% of that achieved by Velcade, PS-1145 did abrogate the protective effect of interleukin-6 against dexamethasone-induced apoptosis [18].

Abrogation of cell survival

A critical component of tumorigenesis is the ability of the cancer cell to evade the induction of apoptosis. This ability to evade apoptosis may arise through defects in the apoptotic pathway *per se*, for example by over-expression of *bcl-2*, or through the activation of survival pathways capable of overriding the commitment to die.

Constitutive activation of survival pathways has been implicated in the aetiology of a number of cancers. For instance, elevated activity of the PI 3-kinase (phosphoinositide 3-kinase)/Akt survival pathway has been observed in a number of tumours including breast, prostate and lung carcinoma (for a review see [19]). Constitutive activation arises, among other causes, from aberrant activation of PI 3-kinase or Akt (or one of their isoforms), or inactivation (mutation or deletion) of PTEN (phosphatase and tensin homologue deleted on chromosome 10; one of the most commonly mutated proteins in cancers; Figure 3).

The frequency of mutation associated with the PI 3-kinase signal cascade makes targeting this pathway for inhibition an attractive therapeutic strategy for cancer. ProlX Pharmaceuticals have identified a group of four small molecules (DPI analogues) which are capable of either inhibiting PI 3-kinase and Akt or inhibiting Akt alone. Pre-clinical studies are currently underway to determine which analogue gives the best *in vivo* anti-tumour activity and therapeutic index.

UCN-01 is a protein kinase inhibitor that is under development as an anti-cancer therapy. Although UCN-01 was originally isolated from the culture broth of *Streptomyces* sp. as a protein kinase C-selective inhibitor, more recent studies have shown that it inhibits the upstream Akt kinase, 3-phosphoinositide-dependent protein kinase 1 (PDK1) (IC_{50}, 33 nM). This inhibition results in the inability of Akt to provide survival signals, consequently apoptosis ensues [20].

A number of other drugs that also target survival/activation pathways have been developed and are showing efficacy in clinical trials.

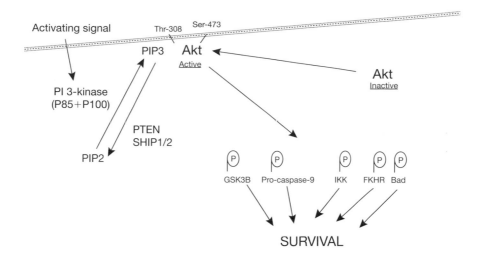

Figure 3. The PI 3-kinase/Akt pathway
This pathway is one of the key survival pathways in transformed cells. Activation of PI 3-kinase causes the generation of PIP3 (PtdInsP$_3$). Levels of PIP3 are regulated by the phosphatases PTEN, SHIP1 and SHIP2 (SH2-containing inositol 5'-phosphatase 1 and 2). PIP3 attracts Akt to the membrane through a pleckstrin homology domain where it can be activated by PDK1 and PDK2 through a dual phosphorylation event. Phosphorylated Akt has a number of anti-apoptotic effects including phosphorylation of GSK3β (glycogen synthase kinase 3β), pro-caspase-9, IKK (inhibitor of NF-κB), Forkhead transcription factors (FKHR) and Bad. Targeting this pathway may prove very efficacious in the treatment of malignant diseases. PIP2, PtdInsP$_2$.

Farnesyl transferase inhibitors

Mutated Ras oncogene-encoding proteins that are constitutively active can induce tumours by activation of Ras/Raf/MEK (mitogen-activated protein kinase/extracellular-signal-regulated kinase kinase) kinase cascade. In human malignancies, Ras mutations are common, having been identified in approx. 30% of cancers. The Ras protein requires farnesylation for activity; therefore inhibition of farnesyl transferase activity is an important therapeutic strategy. Small-molecule inhibitors of farnesylation are showing promise in clinical trials. Presently, two of the most advanced farnesyl transferase inhibitors are Zarnestra and Sarasar, both of which have demonstrated biological and clinical activity against a range of solid tumours. Phase I clinical trials with Zarnestra, for example, yielded clinical responses in ten of the 34 patients (29%), including two complete remissions in adults with refractory or relapsed acute leukaemia, demonstrating that inhibitors of farnesylation may have important clinical anti-leukaemic activity [21]. In a further study, the effect of Zarnestra on multiple myeloma cells was examined and a dose-dependent inhibition of three cell lines tested was reported, together with a significant and time-dependent induction of apoptosis. Moreover, Zarnestra also induced apoptosis in the bone marrow mononuclear cell population of four multiple myeloma patients, being almost restricted to the malignant plasma cells [22].

Tyrosine kinase inhibitors

Protein kinases are the critical mediators of cellular survival signals. Several drugs targeting a variety of protein kinases implicated in survival pathway signalling have recently completed or are currently undergoing clinical trials, and have proven to be very effective in a number of cancers.

Constitutive tyrosine kinase activity of the Bcr–Abl fusion protein (the product of the Philadelphia chromosome) has been established as the causative event in CML (chronic myelogenous leukaemia). Thus the Bcr–Abl tyrosine kinase is an ideal candidate for pharmacological inhibition. Gleevec is an Abl-specific tyrosine kinase inhibitor that, in pre-clinical studies, selectively kills Bcr–Abl-containing cells *in vitro* and *in vivo* [23]. Since the first patient was treated in 1998, approx. 15 000 people worldwide have been treated with Gleevec. In the treatment of chronic-phase CML, Gleevec produces better haematological and cytogenetic responses than the more traditionally used interferon-α, with the majority of patients sustaining these responses. In a phase II study of Gleevec in patients with chronic-phase CML, in whom treatment with interferon had failed, the rates of major and complete cytogenetic responses were 60 and 41%, respectively, compared with 15% for major and 5–7% for complete cytogenetic responses in patients treated with interferon [24].

Currently, other forms of tyrosine kinase inhibition targeting the peptide-binding site rather than the ATP-binding site are being investigated. One reagent, a member of the tyrphostin family termed AG957, and its adamantyl ester (Adaphostin) are currently being investigated as alternative treatments to counter Gleevec resistance.

Another tyrosine kinase inhibitor, PKI166, that inhibits activity of the HER [human EGF (epidermal growth factor) receptor] family, has been shown to be effective in arresting the growth of oral cancer *in vitro* and reduces tumour proliferation in experimental xenograft animal models [25]. In further animal studies, the authors compared the efficacy of PKI166 in EGF+ and EGF− transplanted tumours in nude mice. PKI166 was more effective against EGF+ tumours, either alone or in combination with Paclitaxel, and this was associated with apoptosis of tumour-associated endothelial cells [26].

Receptor-mediated approaches

Death receptors

Upon ligation, death receptors of the TNF receptor superfamily, such as FasR/CD95, mediate apoptosis through the formation of a protein complex known as the DISC (death-inducing signalling complex) [27]. Pro-caspase-8 is recruited to the DISC, where it is processed and released from the complex to activate downstream 'effector' caspases, and ultimately cell death.

The accessibility of cell-surface death receptors makes them attractive targets for apoptotic therapeutics. Initial strategies based on this model focused on either FasL or TNFα; however, due to the acute and severe toxicity associated with administration in animals, the therapeutic usefulness of these particular death receptors ligands was questioned.

A more promising alternative has now emerged based on a third member of the TNF family, TRAIL (TNF-related apoptosis-inducing ligand). TRAIL exhibits a high degree of specificity for tumour cells, owing to the differential tumour-associated expression of its cognate death receptors DR4 and DR5. A number of *in vitro* studies have shown that many tumour cell lines of divergent origins, including cancers of the colon, breast and prostrate, are sensitive to TRAIL-induced apoptosis.

The differential expression of TRAIL receptors on tumour compared with normal cells may be due to a DNA-damage-dependent mechanism. If this is proven to be the case, there is good reason to suspect that TRAIL might also be clinically beneficial in combination therapy with chemotherapeutics presently in use.

Gliomas represent a potential target for this strategy, as both DR4 and DR5 are expressed to some extent in glioma cells, but not in normal brain tissue. Treatment with sub-lethal doses of the DNA-damaging agents cisplatin and etoposide, commonly used to treat gliomas, up-regulate the DR5 receptor but do not induce apoptosis in most of the glioma cells tested. Treatment with soluble TRAIL in combination with the DNA-damaging drugs at sub-lethal doses caused a dramatic and synergistic cell death through TRAIL–receptor interaction and caspase activation *in vitro*, as shown by the fact that neutralizing TRAIL antibodies and caspase inhibition significantly reduced the cytotoxicity of combination therapy. Furthermore, combination treatment with soluble TRAIL and cisplatin resulted in synergistic growth suppression and even regression of established human glioblastoma xenografts in nude mice, as well as inhibition of tumour formation without causing substantial toxicity to the animal [28].

As with FasL and TNFα, there have been safety concerns associated with TRAIL. In contradiction to results obtained in mice, human primary hepatocytes were found to be sensitive to human TRAIL and were induced to undergo rapid apoptosis upon TRAIL exposure. These results yet again raise a question over using death receptors in the clinic [29]. Subsequent studies have, however, indicated that the *in vitro* toxicity against human hepatocytes was related to the polyhistidine-tagged version of TRAIL [30], clearing the way for pre-clinical trials using the native sequence, recombinant soluble ligand.

In a pre-clinical animal study, TRAIL doses given to chimpanzees and cynomolgus monkeys proved to have no hepatic toxicity, even though serum concentrations were as much as 3500-fold higher than those used with the polyhistidine-tagged version *in vitro*. Due to its large size, TRAIL does not penetrate beyond the perivascular space of solid tumours, although this does not seem to affect efficacy, at least not in xenograft models of human cancer [31].

An alternative to TRAIL is the monoclonal antibody TRA-8, which selectively targets the DR5 death receptor. Cancer cells, but not normal cells, primarily express DR5. TRA-8 has been shown to induce apoptosis in glioma cells [32].

Antibodies

Certain monoclonal antibodies, directed against cell-surface differentiation markers or receptors, have been demonstrated to improve the clinical outcome in patients suffering from leukaemia, particularly for those pertaining to B-cells. Rituximab, a chimaeric IgG anti-CD20 monoclonal antibody, was one of the first monoclonal antibodies approved by the FDA in 1997 for treatment of relapsed or refractory, CD20$^+$, B-cell, low-grade or follicular non-Hodgkin's lymphoma. More recently, there have been reports of clinical benefit when Rituximab is used to treat CD20-expressing haematological malignancies, including aggressive non-Hodgkin's lymphoma, mantle cell lymphoma, chronic lymphocytic leukaemia, small lymphocytic lymphoma and Waldenstrom's macroglobulinaemia. Its ability to induce apoptosis in target cells is believed to be due in part to immune-mediated effects, including complement-mediated lysis and antibody-dependent cell-mediated cytotoxicity, and direct effects induced by CD20 ligation.

Conjugation of a cell-specific antibody to a toxin or isotope is being used therapeutically to direct the apoptosis-inducing agent to the particular malignancy, thus limiting the collateral damage associated with chemotherapy. For example, the CD20 antigen is being targeted with Bexxar, a radiolabelled anti-CD20 monoclonal antibody in which the isotope emits β and γ particles. Through a combination of monoclonal antibody plus the DNA-damaging effects of the iodine (^{131}I) radioisotope, malignant cells are induced to undergo apoptosis, while simultaneously, normal tissue has limited exposure. Zevalin is a similar product where the anti-CD20 antibody is the carrier for yttrium-90.

Induction of apoptosis by targeting HLA-DR with monoclonal antibodies is another attractive therapeutic approach being evaluated. Advantage is taken of the fact that the class II HLA-DR receptor has relatively limited expression on non-activated cells, whereas it is more pronounced on malignant cell types. A humanized antibody specific for DR chains, called Remitogen, is in phase II clinical trials for patients with relapsed or refractory grade I, II or III B-cell non-Hodgkin's lymphoma. Despite disappointing results where only one patient in 25 had a clinical response, further investigation is warranted since other investigators have reported pre-clinical responses when Remitogen was given in conjunction with granulocyte colony-stimulating factor [33]. The requirement for granulocyte colony-stimulating factor indicates that apoptosis induced by Remitogen is immune-mediated. However, a potentially more potent humanized HLA-DR antibody has been developed that directly initiates the apoptotic cascade independently of immune effector mechanisms. These antibodies exhibit potent *in vitro* tumoricidal activity on several lymphoma/leukaemia cell lines and on primary leukaemia patient samples.

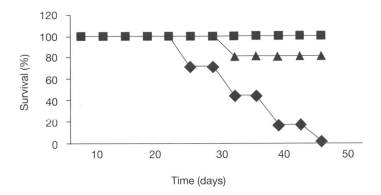

Figure 4. Killing efficiency of anti-HLA-DR antibodies against a xenograft model
SCID mice were injected subcutaneously with a non-Hodgkin's B-cell lymphoma cell line (GRAN-TA-519) prior to treatment with the anti-HLA-DR antibody (1D09C3), which was administered either at 1 mg/mouse on days 5, 7 and 9, subcutaneously (■) or intravenously (▲). Control cells received solvent alone (◆). As can be seen from the graph, administering the anti-HLA-DR antibody protected the animals from death, whereas all control animals died. Reproduced from Nagy, Z.A., Hubner, B., Lohning C., Rauchenberger, R., Reiffert, S., Thomassen-Wolf, E., Zahn, S., Leyer, S., Schier, E.M., Zahradnik, A. et al. (2002) Fully human, HLA-DR-specific monoclonal antibodies efficiently induce programmed cell death of malignant lymphoid cells. *Nat. Med.* **8**, 801–807, with permission from the Nature Publishing Group (www.nature.com/nm/).

Activity has also been reported *in vivo* in xenograft models of non-Hodgkin's lymphoma [34] (Figure 4).

Over-expression of the *HER2/neu* oncogene is a frequent molecular event in multiple human cancers, including breast and ovarian cancer. Patients with cancers where *HER2/neu* is over-expressed have a poorer survival profile. Herceptin, a humanized anti-HER2 monoclonal antibody, has shown significant efficacy in the treatment of HER2-positive metastatic breast cancer. Although the mechanism of action has not been fully defined, it is thought to cause remission by affecting the cell cycle and possibly inducing apoptosis by inhibiting the PI 3-kinase/Akt pathway [35]. Additionally, Herceptin has been shown to enhance the anti-tumour effect of Taxol in combination studies, by allowing effective p34 (CDC2) activation, a necessary step in Taxol-induced apoptosis [36].

Antisense therapy

Modulation of gene expression by introduction of exogenous nucleic acids has been the subject of intense study and development over the past decade. Offering a route to single-gene-directed therapeutics; nucleic acid strategies have the potential to significantly expand the number of potential targets that regulate apoptosis in disease states. Among the approaches investigated thus far are antisense oligonucleotides, ribozymes, DNAzymes, RNA interference and alteration of RNA splicing. As the antisense approach is most developed for apoptosis therapy, this review will only focus on this; however, for a more

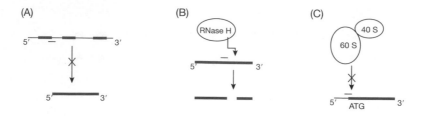

Figure 5. Principal mechanisms of antisense action
Antisense oligonucleotides may inhibit splicing of pre-mRNA (**A**), activate RNase H cleavage of the target mRNA (**B**) or inhibit ribosome binding and translation (**C**).

complete review of these alternate technologies the reader is referred to Opalinska and Gewirtz [37].

Antisense methodology employs the introduction of short synthetic stretches of DNA (oligonucleotides) that are designed to hybridize with specific mRNA strands. The mRNA may then be targeted for RNase-dependent degradation, or, alternatively, the translation of the mRNA may be inhibited. In both cases, synthesis of the target protein is reduced (Figure 5).

Genasense is an 18-mer all-phosphorothioate Bcl-2 antisense oligonucleotide that binds to the first six codons of the Bcl-2 mRNA, and recruits RNAse H, resulting in cleaved message and subsequent down-regulation of Bcl-2 protein and apoptosis [38]. In a pre-clinical animal model, Genasense demonstrated efficacy in SCID mice transplanted with human tumours when delivered by continuous subcutaneous infusion. Induction of apoptosis was further enhanced by co-administration with standard chemotherapeutic agents, indicating that lower Bcl-2 levels decrease the apoptotic threshold of these cells and make them more chemosensitive. This therapy has translated into the clinic in phase I/IIa studies, which demonstrated that in 20 patients, 45% had a clinical response [39]. G-3139 is currently in phase III clinical trials for chronic lymphocytic leukaemia, non-small-cell lung cancer, malignant melanoma and multiple melanoma.

Gene therapy

Gene therapy represents a new therapeutic modality that relies on the transfer of DNA coding for a particular protein into patients. Initially thought of as a treatment for genetic diseases, it is currently being explored for a wide range of acquired disorders, including cancer, cardiovascular diseases, arthritis and neurodegenerative disorders.

One approach for induction of apoptosis in tumours via gene transfer involves transduction of the p53 tumour-suppressor gene directly into the tumour, as exemplified by SCH 58500 for ovarian and Advexin for head and neck cancer. SCH 58500 consists of a recombinant adenoviral vector containing the cloned human wild-type p53 tumour-suppressor gene cDNA under the control of the human cytomegalovirus immediate-early promoter/enhancer element. Preliminary data from phase I/II clinical trials of patients with ovarian cancer suggest that it is well tolerated. Combined with platinum-based therapies, it is associated with a significant reduction of serum CA125 (a tumour marker, used to monitor the status of the ovarian cancer). Long-term follow up of patients revealed a 12–13-month median survival in a heavily pre-treated population with recurrent ovarian cancer. This outcome compares favourably with the 16-month median survival for individuals treated with Paclitaxel at the time of initial recurrence of the disease, and is more than double the 5-month survival seen in patients treated with palliative radiotherapy or who experience Paclitaxel failure [40]. SCH 58500 has now moved on to phase III trials.

The p53 pathway is also the focus the oncolytic viral drug ONYX-015. Rather than over-expressing the p53 gene, ONYX-015 uses a modified adenovirus that is replication-competent in p53-null cells, but not in p53-positive cells. Upon replication in p53-negative cells, the virus lyses the host cell to release its progeny. Since p53 is mutated in 45–70% of all cases of head and neck cancer, ONYX-015 was developed as a tumour-cell-specific therapeutic agent. Moreover, ONYX-015 used in combination with cisplatin and 5-fluorouracil provides enhanced tumour regression in patients [41].

Although there are no cured cases with apoptosis gene therapy, the intensive developments in this area augur well for the future. A number of outstanding issues, principally relating to the limitations of the current vectors employed, still need to be addressed for gene therapy to be an effective apoptotic therapeutic. Chief among these concerns are transduction efficiency, tissue target selectivity and possible oncogenicity or mutagenicity in healthy tissue due to pernicious integration.

Mitochondrion-based therapies

Mitochondrial membrane permeabilization is an often critical event in the process leading to apoptosis. This permeabilization event is, at least in part, under the control of the permeability transition pore complex. Bcl-2 family members and tumour-suppressor proteins from the Bax family interact with the permeability transition pore complex to inhibit and facilitate membrane permeabilization, respectively. Conventional chemotherapeutic agents elicit mitochondrial permeabilization in an indirect fashion by induction of endogenous effectors that are involved in the physiological control of apoptosis. An increasing number of experimental anti-cancer drugs are being discovered that act directly on mitochondrial membranes through changes in cellular redox potential due to enhanced generation of reactive oxygen species. Reactive oxygen species are generated during the production of ATP by aerobic metabolism in the mitochondria and are typically oxide (O_2^-) and hydroxyl ($OH^\bullet$) radicals. Excessive amounts of these reactive species can start a lethal chain reaction such as oxidation of membrane phospholipids that together with other structural disablement can affect cellular integrity and survival. Such agents may induce apoptosis under circumstances in which conventional drugs fail to act due to mutation/deletions in key apoptotic machinery. However, stabilization of the mitochondrial membrane by anti-apoptotic Bcl-2-like proteins reduces the cytotoxic potential of most of these drugs and therefore novel therapeutics that are independent of Bcl-2 status are also required.

Photodynamic therapy

Photodynamic therapy employs the combination of a photosensitizing chemical and light to initiate apoptosis. Based on this concept, photosensitizers have been targeted to neoplastic tissue and key subcellular organelles, such as the mitochondria. Activation with the appropriate wavelength of light generates reactive oxygen species that harm membranes and reduce the activity of membrane-associated enzymes, such as succinic dehydrogenase in the mitochondria, inducing apoptosis in the diseased tissue. Photofrin photodynamic therapy has received FDA approval for the palliative treatment of totally and partially obstructing oesophageal malignancies and for early-stage microinvasive lung cancer. Because Photofrin has undesirable side effects of skin photosensitization, the next generation of photosensitizers is being developed. Photodynamic therapy using the second-generation photosensitizer phthalocyanine 4 (Pc4) causes mitochondrial swelling and damage, subsequent to apoptosis induction through the release of cytochrome c into the cytosol. Furthermore, in a mouse tumour model in which RIF-1 tumours were transplanted to C3H/HeN mice, it was shown that Pc4 was more efficacious than Photofrin, as determined by tumour regression [42].

Retinoic acid

Retinal (vitamin A) and its naturally occurring and synthetic analogues (retinoids) exert profound effects on development, cellular differentiation and apoptosis. Due to their strong differentiative and anti-proliferative activities, they have attracted interest as potential anti-cancer and cancer prevention agents. The paradigm for retinoic-acid-mediated cancer resolution is APL (acute promyelocytic leukaemia). The causative event in APL is the formation of a PML (promyelocytic leukaemia)–RARα (retinoic acid receptor α) fusion protein, which represses signalling and consequently terminal differentiation. High non-physiological doses of all-*trans* retenoic acid overcome this block, restore normal signalling and induce terminal differentiation with subsequent apoptosis leading to remission [43]. The anti-cancer effects of naturally occurring retinoids are mainly mediated by their nuclear receptors, the RARs and the RXRs (retinoid X receptors). Retinoids promote apoptosis in breast and lung cancer cells mainly through RARβ. RARβ functions as a tumour-suppressor gene in carcinogenesis and is either activated directly (hormone-independent) or indirectly induced via RAR/RXR heterodimers (hormone-independent) [44].

The clinical use of natural retinoids has been limited due to both the rapid emergence of retinoid resistance in cancer cells and the toxicity associated with these molecules. These limitations are currently being challenged by the introduction of novel synthetic retinoids such as fenretinide and AHPN (6-[3-(1-adamantyl)-4-hydroxyphenyl]-2-naphthalene carboxylic acid). Both of these synthetic retinoids can induce apoptosis in all-*trans* retenoic-acid-resistant models, suggesting that their induction of apoptosis has an RAR/RXR-independent mode of action. This RAR/RXR-independent effect has been shown to be due to mitochondrial dysfunction via the induction of mitochondrial permeability transition [45,46]. Both types of retinoid induce reactive oxygen species in cells and this is key to synergism with chemotherapeutic drugs [47]. In phase III trials of 2972 patients, fenretinide was associated with a 35% relative reduction in the incidence of contralateral and ipsilateral breast cancers in pre-menopausal women (mostly 50 years of age), whereas a trend to an increased number of contralateral cancers was observed in post-menopausal women [48].

Arsenic trioxide (As_2O_3)

As_2O_3 inhibits the growth and differentiation of a variety of tumour types, including myeloid leukaemias, myeloma, lymphoid leukaemia and solid tumours including prostate, ovarian and neuroblastoma cells. The therapeutic effect against APL is mediated by induction of apoptosis, even in patients resistant to all-*trans* retinoic acid [49]. In studies using the APL cell line (NB4), As_2O_3 was shown to induce apoptosis by down-regulating Bcl-2 together with inhibition of other proteins that are specific to this haematological disease (PML and PML–RARα). Various groups have demonstrated that the mechanism behind As_2O_3-mediated apoptosis is through the induction of reactive oxygen species. Work by Woo et al. [50] demonstrated that, in HeLa cells, As_2O_3 increased cellular content of reactive oxygen species, especially H_2O_2. Moreover, treatment of the cells with either the antioxidant *N*-acetyl-L-cysteine or catalase suppressed As_2O_3-induced apoptosis. Indeed, other researchers have demonstrated that the inherent ability of leukaemic cells to generate reactive oxygen species ultimately determines the sensitivity of cells to undergo apoptosis upon exposure to this reagent [51].

As_2O_3 has an impressive success rate in the clinic, despite adverse side effects that include abdominal discomfort, nausea, vomiting, headache, fatigue, skin changes and fluid accumulation (most of which were considered mild and resolved after therapy was completed). In a multicentre study in the United States, 40 patients experiencing either a first (*n*=21) or at least a second (*n*=19) relapse of APL were treated. Some 86% of the patients converted from positive to negative for PML/RARα. The 18-month overall survival and relapse-free survival estimates were 66 and 56% respectively [52]. The FDA has approved Trisenox (As_2O_3) for the treatment of patients with APL who have not responded to, or have relapsed following the use of, all-*trans* retinoic acid and anthracycline-based chemotherapy, which is considered first-line therapy.

Future perspectives

Understanding the apoptotic pathway has the potential to revolutionize clinical intervention and signal the end of non-specific chemotherapy, thus reducing morbidity and mortality. In this brief overview, it is envisaged that the reader gains an appreciation of the ever-increasing number of potential routes to the regulation of apoptosis. However, therapeutics based on this principle that are either on the market or currently undergoing clinical trials represent only the tip of the proverbial iceberg. As this review illustrates, there is an expanding array of ways to regulate expression of both aberrant gene expression and protein activity. Where there will be a limit to apoptosis therapeutics is in the identification of causative novel targets suitable for development as drug targets. It is therefore hoped that, with current genomic and proteomics capabilities together with computer-simulation models of key cellular pathways, which can be substantiated by high-throughput target validation, the apoptotic pathways will be unravelled to discover novel targets, which in turn will translate into novel, exciting and effective therapeutics.

Summary

- *Dysregulation of apoptosis has been implicated in the vast majority of diseases affecting humankind today.*
- *Historically, successful cancer and inflammatory therapies have inadvertently been apoptosis-modifying drugs.*
- *A number of novel therapeutic approaches are currently being examined. These approaches include traditional small molecules, but also encompass more novel approaches including antisense methodologies and antibodies.*
- *Some therapeutics already in the clinic are showing promising efficacy, particularly in the fight against cancer.*
- *Future-generation apoptosis-modifying drugs will have the potential to target the diseased cell while ignoring normal cells, and reducing and, in some cases, eliminating the side effects of therapy.*

We thank our colleagues at EiRx Therapeutics for their continued support. We also thank Professor Finbarr E. Cotter for his advice and Dr Dyan Sheehan for her critical reading of this manuscript.

References

1. Endres, M., Namura, S., Shimizu-Sasamata, M., Waeber, C., Zhang, L., Gomez-Isla, T., Hyman, B.T. & Moskowitz, M.A. (1998) Attenuation of delayed neuronal death after mild focal ischemia in mice by inhibition of the caspase family. *J. Cereb. Blood Flow Metab.* **18**, 238–247

2. Cursio, R., Gugenheim, J., Ricci, J.E., Crenesse, D., Rostagno, P., Maulon, L., Saint-Paul, M.C., Ferrua, B. & Auberger, A.P. (1999) A caspase inhibitor fully protects rats against lethal normothermic liver ischemia by inhibition of liver apoptosis. *FASEB J.* **13**, 253–261

3. Mocanu, M.M., Baxter, G.F. & Yellon, D.M. (2000) Caspase inhibition and limitation of myocardial infarct size: protection against lethal reperfusion injury. *Br. J. Pharmacol.* **130**, 197–200

4. Hotchkiss, R.S., Chang, K.C., Swanson, P.E., Tinsley, K.W., Hui, J.J., Klender, P., Xanthoudakis, S., Roy, S., Black, C., Grimm, E. et al. (2000) Caspase inhibitors improve survival in sepsis: a critical role of the lymphocyte. *Nat. Immunol.* **1**, 496–501

5. Leist, M. & Jaattela M. (2001) Four deaths and a funeral: from caspases to alternative mechanisms. *Nat. Rev. Mol. Cell Biol.* **2**, 589–598

6. Hu, S., Snipas, S.J., Vincenz, C., Salvesen, G. & Dixit, V.M. (1998) Caspase-14 is a novel developmentally regulated protease. *J. Biol. Chem.* **273**, 29648–29653

7. de Bilbao, F., Guarin, E., Nef, P., Vallet, P., Giannakopoulos, P. & Dubois-Dauphin M. (1999) Postnatal distribution of cpp32/caspase 3 mRNA in the mouse central nervous system: an in situ hybridization study. *J. Comp. Neurol.* **409**, 339–357

8. Donovan, M. & Cotter, T.G. (2002) Caspase-independent photoreceptor apoptosis *in vivo* and differential expression of apoptotic protease activating factor-1 and caspase-3 during retinal development. *Cell Death Differ.* **9**, 1220–1231

9. Reed, J.C. (1999) Mechanisms of apoptosis avoidance in cancer. *Curr. Opin. Oncol.* **11**, 68–75

10. Wang, J.L., Liu, D., Zhang, Z.J., Shan, S., Han, X., Srinivasula, S.M., Croce, C.M., Alnemri, E.S. & Huang, Z. (2000) Structure-based discovery of an organic compound that binds Bcl-2 protein and induces apoptosis of tumor cells. *Proc. Natl. Acad. Sci. U.S.A.* **97**, 7124–7129

11. Komarov, P.G., Komarova, E.A., Kondratov, R.V., Christov-Tselkov, K., Coon, J.S., Chernov, M.V. & Gudkov, A.V. (1999) A chemical inhibitor of p53 that protects mice from the side effects of cancer therapy. *Science* **285**, 1733–1737

12. Duan, W., Zhu, X., Ladenheim, B., Yu, Q.S., Guo, Z., Oyler, J., Cutler, R.G., Cadet, J.L., Greig, N.H. & Mattson, M.P. (2002) p53 inhibitors preserve dopamine neurons and motor function in experimental parkinsonism. *Ann. Neurol.* **52**, 597–606

13. Keer, H.N., Reid, T. & Sreedharan, S. (2002) Pivanex activity in refractory non-small cell lung cancer, a phase II study. In 38th Annual Meeting of the American Society of Clinical Oncology (ASCO), abstract 1253

14. Batova, A., Shao, L.E., Diccianni, M.B., Yu, A.L., Tanaka, T., Rephaeli, A., Nudelman, A. & Yu, J. (2002) The histone deacetylase inhibitor AN-9 has selective toxicity to acute leukemia and drug-resistant primary leukemia and cancer cell lines. *Blood* **100**, 3319–3324

15. Hideshima, T., Richardson, P., Chauhan, D., Palombella, V.J., Elliott, P.J., Adams, J. & Anderson, K.C. (2001) The proteasome inhibitor PS-341 inhibits growth, induces apoptosis, and overcomes drug resistance in human multiple myeloma cells. *Cancer Res.* **61**, 3071–3076

16. LeBlanc, R., Catley, L.P., Hideshima, T., Lentzsch, S., Mitsiades, C.S., Mitsiades, N., Neuberg, D., Goloubeva, O., Pien, C.S., Adams, J. et al. (2002) Proteasome inhibitor PS-341 inhibits human myeloma cell growth *in vivo* and prolongs survival in a murine model. *Cancer Res.* **62**, 4996–5000

17. Hideshima, T. & Anderson, K.C. (2002) Molecular mechanisms of novel therapeutic approaches for multiple myeloma. *Nat. Rev. Cancer* **2**, 927–937

18. Hideshima, T., Chauhan, D., Richardson, P., Mitsiades, C., Mitsiades, N., Hayashi, T., Munshi, N., Dang, L., Castro, A., Palombella, V. et al. (2002) NF-kappa B as a therapeutic target in multiple myeloma. *J. Biol. Chem.* **277**, 16639–16647

19. Vivanco, I. & Sawyers, C.L. (2002) The phosphatidylinositol 3-kinase-AKT pathway in human cancer. *Nat. Rev. Cancer* **2**, 489–501

20. Sato, S., Fujita, N. & Tsuruo T. (2002) Interference with PDKI-Akt survival signaling pathway by UCN-0I (7-hydroxystaurosporine). *Oncogene* **21**, 1727–1738

21. Karp, J.E., Lancet, J.E., Kaufmann, S.H., End, D.W., Wright, J.J., Bol, K., Horak, I., Tidwell, M.L., Liesveld, J., Kottke, T.J. et al. (2001) Clinical and biologic activity of the farnesyltransferase inhibitor R115777 in adults with refractory and relapsed acute leukemias: a phase I clinical-laboratory correlative trial. *Blood* **97**, 3361–3369

22. Le Gouill, S., Pellat-Deceunynck, C., Harousseau, J.L., Rapp, M.J., Robillard, N., Bataille, R. & Amiot, M. (2002) Farnesyl transferase inhibitor R115777 induces apoptosis of human myeloma cells. *Leukemia* **16**, 1664–1667

23. Mauro, M.J., O'Dwyer, M.E. & Druker, B.J. (2001) ST1571, a tyrosine kinase inhibitor for the treatment of chronic myelogenous leukemia: validating the promise of molecularly targeted therapy. *Cancer Chemother. Pharmacol.* **48**, suppl. 1, S77–S78

24. Kantarjian, H., Sawyers, C., Hochhaus, A., Guilhot, F., Schiffer, C., Gambacorti-Passerini, C., Niederwieser, D., Resta, D., Capdeville, R., Zoellner, U. et al. (2002) Hematologic and cytogenetic responses to imatinib mesylate in chronic myelogenous leukemia. *N. Engl. J. Med.* **346**, 645–652

25. Myers, J.N., Holsinger, F.C., Bekele, B.N., Li, E., Jasser, S.A., Killion, J.J. & Fidler, I.J. (2002) Targeted molecular therapy for oral cancer with epidermal growth factor receptor blockade: a preliminary report. *Arch. Otolaryngol. Head Neck Surg.* **128**, 875–879

26. Baker, C.H., Kedar, D., McCarty, M.F., Tsan, R., Weber, K.L., Bucana, C.D. & Fidler, I.J. (2002) Blockade of epidermal growth factor receptor signaling on tumor cells and tumor-associated endothelial cells for therapy of human carcinomas. *Am. J. Pathol.* **161**, 929–938

27. Medema, J.P., Scaffidi, C., Kischkel, F.C., Shevchenko, A., Mann, M., Krammer, P.H. & Peter, M.E. (1997) FLICE is activated by association with the CD95 death-inducing signaling complex (DISC). *EMBO J.* **16**, 2794–2804

28. Nagane, M., Pan, G., Weddle, J.J., Dixit, V.M., Cavenee, W.K. & Huang, H.J. (2000). Increased death receptor 5 expression by chemotherapeutic agents in human gliomas causes synergistic cytotoxicity with tumor necrosis factor-related apoptosis-inducing ligand *in vitro* and *in vivo*. *Cancer Res.* **60**, 847–853

29. Jo, M., Kim, T.H., Seol, D.W., Esplen, J.E., Dorko, K., Billiar, T.R. & Strom, S.C. (2000) Apoptosis induced in normal human hepatocytes by tumor necrosis factor-related apoptosis-inducing ligand. *Nat. Med.* **6**, 564–567

30. Lawrence, D., Shahrokh, Z., Marsters, S., Achilles, K., Shih, D., Mounho, B., Hillan, K., Totpal, K., DeForge, L., Schow, P. et al. (2001) Differential hepatocyte toxicity of recombinant Apo2L/TRAIL versions. *Nat. Med.* **7**, 383–385

31. Kelley, S.K., Harris, L.A., Xie, D., Deforge, L., Totpal, K., Bussiere, J. & Fox, J.A. (2001) Preclinical studies to predict the disposition of Apo2L/tumor necrosis factor-related apoptosis-inducing ligand in humans: characterization of *in vivo* efficacy, pharmacokinetics, and safety. *J. Pharmacol. Exp. Ther.* **299**, 31–38

32. Choi, C., Kutsch, O., Park, J., Zhou, T., Seol, D.W. & Benveniste, E.N. (2002) Tumor necrosis factor-related apoptosis-inducing ligand induces caspase-dependent interleukin-8 expression and apoptosis in human astroglioma cells. *Mol. Cell. Biol.* **22**, 724–736

33. Stockmeyer, B., Schiller, M., Repp, R., Lorenz, H.M., Kalden, J.R., Gramatzki, M. & Valerius T. (2002) Enhanced killing of B lymphoma cells by granulocyte colony-stimulating factor-primed effector cells and HuID10 – a humanized human leucocyte antigen DR antibody. *Br. J. Haematol.* **118**, 959–967

34. Nagy, Z.A., Hubner, B., Lohning, C., Rauchenberger, R., Reiffert, S., Thomassen-Wolf, E., Zahn, S., Leyer, S., Schier, E.M., Zahradnik, A. et al. (2002) Fully human, HLA-DR-specific monoclonal antibodies efficiently induce programmed death of malignant lymphoid cells. *Nat. Med.* **8**, 801–807

35. Yakes, F.M., Chinratanalab, W., Ritter, C.A., King, W., Seelig, S. & Arteaga, C.L. (2002) Herceptin-induced inhibition of phosphatidylinositol-3 kinase and Akt Is required for antibody-mediated effects on p27, cyclin D1, and antitumor action. *Cancer Res.* **62**, 4132–4141